GOLDPANNING
IN THE CARIBOO
A PROSPECTOR'S TREASURE TRAIL TO CREEKS OF GOLD

by
Jim Lewis & Charles Hart

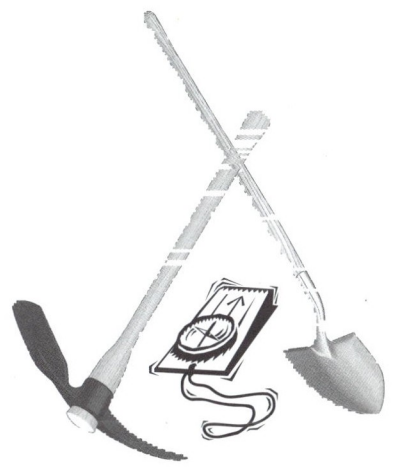

Heritage House
Publishing Company Ltd.

CANADIAN CATALOGUING IN PUBLICATION DATA
Lewis, Jim, 1939-
Goldpanning in the Cariboo

(Creeks of gold series)
ISBN 1-895811-33-3

1. Cariboo (B.C.: Regional district)-Guidebooks.
2. Prospecting-British Columbia-Cariboo Region-Guidebooks.
3. Cariboo (B.C.: Regional district)-Gold discoveries.
I. Hart, Charles, 1954-II. Title. III. Series

FC3845.C3L48 1997 917.11'75 C97-910136-0
F1089.C3M8 1997

First Edition 1997

Heritage House wishes to acknowledge the support of Heritage Canada, the British Columbia Arts Council and the Cultural Services Branch of the Ministry of Small Business, Tourism and Culture. In addition, we encourage our readership to support the BC Archives and Records Service (BCARS) and other institutions acknowledged on our photo credits. They help make publications such as this possible.

Front Cover Photo: Mark Kaarremaa
Back Cover & Jim Lewis Photos: Charles Hart
Cover & Book Design, Layout & Maps: Catherine Mack, Cairn Consulting

I Met an Ancient Pioneer © 1972 Katie Kidwell, used by permission
Go for the Gold © 1984 Jean Ann Robitaille, used by permission
The Mighty Fraser © 1978 Roger Koe (Soulkeeper Music Publishing), used by permission

HERITAGE HOUSE PUBLISHING COMPANY LTD.
Unit #8 - 17921 55th Ave., Surrey, BC V3S 6C4

Printed in Canada by Friesens Printers

ᴀCKNOWLEDGEMENTS

Thanks are in order to several people for invaluable support and assistance in the publication of this book.

To Rodger Touchie at Heritage House Publishing Company Ltd., for his insights, contributions, incisive editing and good humour.

To Cathy Mack for helpful criticism, funky maps and overall dedication to design.

To Katie Kidwell (The Cariboo Song Rider) for I Met an Ancient Pioneer, and for prompt e-mail replies to a stranger's requests.

To Jean Ann Robitaille (Go for the Gold) and Roger Koe (The Mighty Fraser) for use of their song lyrics.

To the Ministry of Employment and Investment (Carron Berkes) for permission to quote from the Ministry of Mines Annual Reports (1874 - 1990).

And finally, a note of thanks to John Williams, button accordionist extraordinaire, whose tasteful tunes helped bounce this project along.

I MET AN ANCIENT PIONEER

I met an ancient pioneer
A miner once he did;
Yep, kept his house with nothen' more
Then a pack sack and a pick.

He tells me this between lean breath.
His eyes a merry squint -
"Ate nothen' then but beans and lard
And sometimes a couple twigs."

"Yep, remember when I struck it rich;
Was walkin' through the town,
Tripped and fell - forgot to YELL
And fell flat on the ground!"

"Well I cussed them boots (upside down!)
When I saw me teeth - layin' around!
Then someone screams and I am told:
me teeth are stuck with powdered gold..."

He coughs and spats (and assures me that)
He wasn't very hurt.
Then he tells me with his toothless grin:
"It's how I hit pay dirt!"

©Katie Kidwell, 1972

CONTENTS

PUBLISHER'S FOREWORD

There is a distinct parallel between the world of prospecting and that of book publishing. The sheer pleasure of finding a nugget amid the pebbles of a Cariboo creek can provide a thrilling and memorable moment. Likewise, the discovery of a manuscript or an author's book proposal, complete with its own hidden treasures, can excite even the most seasoned of word miners.

The words that Jim Lewis first sent to this publisher were written in pen and ink with a steady hand and a sense of conviction - an antiquated beginning in this age of computers and spin doctors. Every creekbed has its own "colour" and it didn't take long to see that the hand-written words on Jim Lewis's lined paper glittered in their own unique way.

Underlying Jim's theme and his overall message, one simple fact shone through. He knew where gold was and he wanted to share his knowledge. This is not a common human trait and demanded further investigation. Jim Lewis turned out to be a man who represented both a way of frontier life and a simple no-strings-attached interest in encouraging others, especially young people, to get out and experience the joy of prospecting.

There was only one problem. When it came right down to it, Jim Lewis was an intricate part of any placer mining story he started to tell. But Jim was not a man to spend much time talking about himself. Briefly, the project floundered on its own sandbar.

Then in a serendipitous turn of events, a pair of unlikely urban prospectors seeking a client, entered the picture and the publisher's office. Business partners Charles Hart and Catherine Mack, writer/editor and graphic designer respectively, brought new dimensions to the challenge at hand. Chuck, with the soul of an Irish musician, and Cath, with roots in a Canadian gold-mining community, proved an ideal match for Jim Lewis and his goldpan.

In ensuing months, over home-cooked meals, along gold-bearing creek beds, or shivering in the outbuilding retreat where Jim keeps his mining library, Chuck Hart absorbed the words and the character of the man who had become his co-author.

The blend of these personalities now manifests itself in the first of a series of books designed to aid both the occasional and the serious prospector throughout B.C.'s many gold-producing regions. To set novice prospectors on the right course, Section One of this book covers both the basic essentials of goldpanning and the bureaucracy that protects individual interests by documenting claims and discoveries.

Section Two hones in on the motherlode and features specific recommendations on gold-bearing sites along six major Cariboo creeks that Jim Lewis has personally prospected. These watersheds and the many tributaries and gulches along their path are still laden with gold. With this book, a twenty-dollar gold pan and a packsack of enthusiasm, you're well on your way to finding some of it.

Rodger Touchie

Jim Lewis
demonstrates
the art of panning

Below, new
prospectors
are born with
the discovery
of 'colour'

*T*HE ESSENTIALS

Nothing lights up the human imagination quite like the flash of gold. Perhaps it's because the great stories of bonanza strikes and glorious riches are rather like fairy tales. With a bit of luck, we, too, might make a rare find.

On the surface, this notion may appear little more than pleasant illusion. From British Columbia's gold history, we tend to infer that the glittering streams of yester-year must now be played out. But a deeper look - in the ground and in the mining records - reveals that untold facets of discovery, now as ever, await the curious wanderer. And the beauty of it all is this: while meandering through B.C.'s splendid valleys you can dabble away in the gold creeks without anything more sophisticated than a round, plastic pan. With a growing sense of the golden opportunities that still exist, you may even decide to go further and stake a claim. That's a relatively simple undertaking too, and we'll look into the details a bit later on.

Compared to today's recreational prospector, the province's early gold-seekers weren't anywhere near so fortunate. Many had to wager everything on a chance at their dreams.

The brilliant metal was first mined in B.C. when the province was a new English colony in the late 1850s. Tens of thousands of fortune hunters made their way north from California, following the mighty Fraser River to the gold fields. When the Cariboo gold rush was declining in the late 1880s, gold was still selling for $16 an ounce. From then on its value was fixed at various levels, until 1971, when the international price was allowed to float according to demand.

With prices in recent years as high as $475 an ounce, it's easy to see why gold has retained its lustre, and why it continues to lure prospectors to remote and uncharted places. We'll let these adventurers pursue their more distant dreams, however, for theirs is not the road travelled in this guide.

Rather, the Creeks of Gold series points toward accessible sites, documented finds, pleasant experience and the ultimate thrill of discovering gold. Hikers, campers, families - anyone who samples the creeks in this guide - will quickly see proof that the best places to search are the very areas which produced gold in the past. Here your chances of success are far greater than wandering about in virgin areas with no history of discovery. For despite all the gold recovered in B.C. over the years, there are a number of good reasons why there's plenty still waiting to be found - as you'll see.

In the end, with recreational panning you never lose. No matter what unfolds - be it a day of exciting discovery or a day of peaceful exploration - either is a treasure. Assuming you are here more for excitement than peace, the rest of this chapter covers what you need to know.

1. **How to select and survey a creek**

2. **Techniques for recovering gold**

3. **Rules of safe prospecting**

4. **Carrying the right equipment**

5. **The prospector's lingo**

1. WHERE TO SEARCH

Every year spring torrents scrape more gold from mineral veins in the high country, carrying flakes and nuggets downstream. Because it is so dense, gold sinks to the streambed in predictable locations. Easily distinguished from other minerals by its weight and brilliant colour, gold gradually works its way through surface gravels and settles on bedrock. One of the most critical aspects of prospecting is developing an eye for the places where gold tends to accumulate.

SURVEYING A CREEK

The most basic rule of finding gold is to look for areas where natural obstructions slow the current, allowing heavier material to sink to the bottom. The swifter the flow, the less likely there'll be gold deposits. The best areas are easiest to spot during spring run-off or after heavy rainfall.

On straight-flowing creeks look for large rocks, sand and gravel bars or uprooted trees - anything that interrupts the water's flow. Iron often originates from the same mineral veins as gold and, being almost as

SURVEY THE CREEK

Flow is faster in these areas

GOLD

GOLD

WATCH FOR OBSTRUCTIONS IN THE FLOW

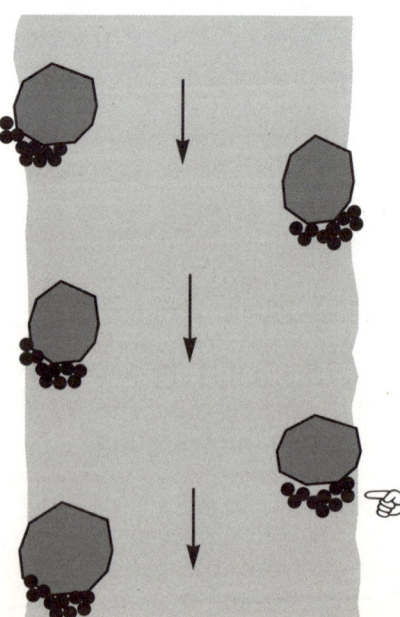

Large rocks like these act as natural sluices to trap gold.

Gold settles on the down-side of boulders and tree trunks.

heavy, tends to accumulate in similar places. Iron pyrites and black sand (concentrated iron) are gold indicators. **Any black sand you find should be panned.** Iron pyrites? Well, they're the infamous fool's gold. Fortunately they can be distinguished from the real McCoy by their sharp edges and brittle texture. Apply pressure with a knife blade and iron pyrites shatter.

Gold is much softer and will become rounded after travelling a short distance downstream. If you do happen to stumble upon rough nuggets, catch your breath and take a good look around. Assuming they haven't tumbled from someone else's poke, chances are you've struck pay dirt close to the source!

SEARCH TIPS

1. Gold is heavier than sand and gravel: it works its way to bedrock.

2. Focus on areas where creeks slow down.

3. Explore the stream bed behind large rocks. Gold accumulates on the downstream side.

4. Check under uprooted trees, the creek bed may be exposed.

5. Watch for iron pyrites and black sand.

LOCATIONS

British Columbia's rich gold history records spectacular discoveries in many areas of the province - Atlin, Boundry and Big Bend, Cassiar, the East and West Kootenays, the Fraser River and its tributaries, the Similkameen, the Thompson River and its tributaries, the Okanagan, Omenica, Vancouver Island, the Queen Charlottes, the McLeod River north of Prince George. The heaviest concentrations have been found in the Cariboo region: in Barkerville, Quesnel Forks, Likely, Horsefly, Quesnel Lake, the Quesnel River area and the Fraser River.

A geological band of gold runs from Alaska and the Yukon Territory right through the heart of British Columbia. These extensive deposits, distributed mainly by the action of glaciers and erosion, continue to shift and evolve through natural forces.

GLACIERS

Glacial action during the retreat of the last ice age, some 15,000 years ago, sprinkled new deposits of gold all along B.C.'s creeks and rivers. Scraping the mountain peaks and carving out valleys, huge blankets of flowing ice ground everything in their path and cast off gold and other minerals in their wake.

Bear in mind that the flow of creeks is ever changing. Earth tremors, landslides and other natural forces can dramatically alter a creek's course, leaving a dry channel where mineral-laden water used to flow. An understanding of these channels, called glacial channels and pre-glacial channels, is an important part of the prospecting mystique.

ANCIENT CHANNELS

Remember: **old channels don't always follow the course of an existing stream, but are often rich sources of gold.** In many cases, these channels have never been mined!

When panning, always be alert for changes in gold content from one spot to the next. If a promising area suddenly turns barren, it's likely an old channel traverses the creek and you've passed beyond it, losing the gold's trail. Go back and carefully rework the area. Check the banks above the water line.

Old channels are often wide and, like surface creeks and streams, tend to meander along crooks and bends. As a general rule, **glacial channels are richer and tend to run in a north-west and south-east direction,** with smaller streams flowing in from the sides.

LOCATING CHANNELS

A visual survey of creek banks is easy to do and may reveal the path of older channels. Climb up to level land, lie flat on your stomach and look for even the slightest hump or depression in the ground. Scan the terrain from both sides of the creek. If there are any irregularities, sample the banks on both sides of the creek at that location, above the high water mark. Although primitive, this approach is as effective today as it was when prospectors were first exploring B.C.'s creeks.

EROSION

Since erosion continues to deposit gold in new locations, panning expeditions can be especially rewarding after a hard winter. When snow and ice melt, huge volumes of water rush down from the headwaters. This annual runoff rakes away boulders, trees, gravel and topsoil, grinding creek banks and scouring stream beds. Along with the flood, gold is swept farther downstream - sometimes settling many miles from its original source.

Gold is unlikely to settle in a section of creek that flows over exposed bedrock. **Always check gravel banks**, though, at creek level and higher up, as well as any cracks and crevices in the rock. The survey tips described earlier will help you locate the best prospects. If you can, go to high ground and try to visualize the creek's flow during high runoff.

2. METHODS OF RECOVERY

Patience is possibly the single most important trait necessary for good prospecting. Recovering gold is a methodical process and should not be rushed.

THE GOLD PAN

For those travelling light, this is the only way to go. The most popular model is the plastic pan with built-in riffles - little ridges that help trap the gold.

These pans have replaced the traditional metal pan. They are lightweight, acid-resistant, do not corrode and are strong enough to withstand rough treatment. Green is the best colour - it readily displays black sand and is ideal for spotting fine gold. Pans range in size from 6 to 18 inches in diameter, the most common being the 12- to 14-inch models. These are easy to handle and cost less than $20.

PANNING TECHNIQUE

Always work your way upstream to avoid disturbing unpanned waters. Sample the creek every twenty or thirty feet, watching for any sudden changes that might announce the presence of an old channel.

Fill the pan with creek-bed gravel to the top riffle, or two-thirds full. Submerge the pan, mix and knead its contents with your hands, being sure to break lumps of clay or soil to free any gold trapped inside. Continue mixing in water and some of the finest material will wash away. Pick out the surface rocks and pebbles and discard them.

Now you're ready to begin in earnest. Keeping the riffled portion away from you, resubmerge the pan then raise the rim to just below the water's surface. Tilt the pan slightly away so the nearside rim is just above water level. Shake the pan from side to side with a light circular motion. This technique keeps the lighter material in suspension and works it out of the pan, leaving gold and other heavy materials around the pan's bottom edges.

Every now and then, lift the pan clear of the creek and shake it vigorously, using the same circular motion to concentrate the gold and heavy sands. With your thumb, again scrape away pebbles and fine materials that rise to the top. Continue panning until only the heaviest material remains. The whole operation should take

1920's prospectors work the Cariboo River with metal pans

Right, good technique involves tilting the pan and swirling it to concentrate the heaviest materials on the bottom

about five or ten minutes (you'll get faster with experience). Be careful to ensure that no gold washes away in the final stages. If you have a tub or pail handy, use it instead of the stream to finish panning. Coarse gold can now be removed with tweezers. Since gold does not respond to magnetism, a magnet is ideal for extracting any black sand or iron pyrites. The remaining material should be dried. The finest gold can be recovered by carefully blowing the dried sand away or, by using mercury.

THE LAST STEP

Purifying the gold residue takes some ingenuity. Here are two simple techniques that have emerged over the years.

a) Mercury and a Chamois Cloth

Placed in a gold pan along with the sand tailings, mercury will amalgamate fine gold, absorbing the tiny particles into a ball. The gold can then be extracted by squeezing the mercury through a chamois cloth.

b) Mercury and a Baked Potato

Alternatively, the trusty potato can be pressed into service. Cut a large potato in two and hollow out one half for the amalgam. Rejoin the halves and cover the whole potato with aluminum foil. After baking it thoroughly, remove the foil and squeeze the potato to drain the mercury. (Although it may be tempting, don't eat the potato!)

The remaining gold can be further purified by placing it in a metal bowl and applying heat, or by using a retort.

ABOUT MERCURY

Mercury is a rather noxious substance that needs to be handled with care. Fortunately, only a small quantity (one or two ounces) is needed to amalgamate fine gold. These days, it's not that easy to find: mining or laboratory suppliers are your best bets.

Prospecting High Banks

Gold will be deposited high up on creek banks during spring runoff. In dry season these areas may be well above the water-line and hard to reach. Here's a simple method for prospecting a high bank when you plan to return some time later.

Look for a small, dry ravine - one that sees water only during high runoff. Clean out the lower section of this gully, removing rocks, brush and limbs. Lay a sheet of plastic, 15-20 feet wide, about 50 feet up the ravine from the main stream. Dig the sheet in at the top to hold it in place, then secure it with rocks along the top and sides. Lay a series of bigger rocks at the bottom to trap material that washes down.

The following spring, or after heavy runoff, pan the material that has collected. This 'float' may contain quartz bearing gold or other signs that the higher banks would be worth closer inspection. (When you're done, be sure to pack the plastic out with you.)

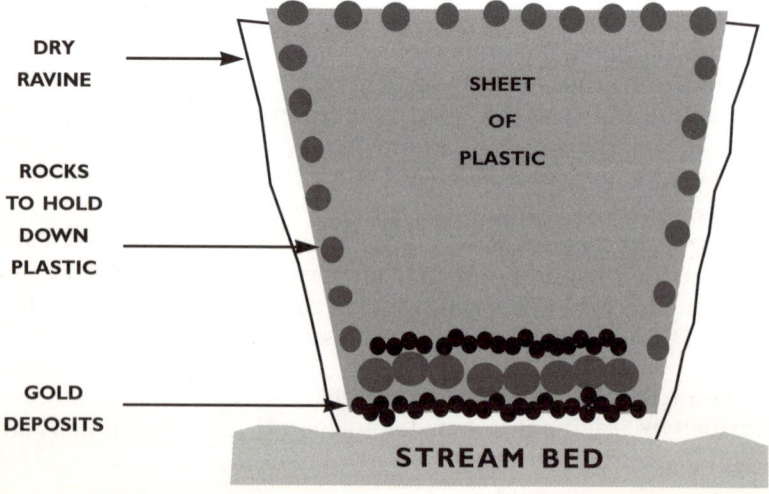

DRY RAVINE

SHEET OF PLASTIC

ROCKS TO HOLD DOWN PLASTIC

GOLD DEPOSITS

STREAM BED

OTHER METHODS

Unless you obtain a free miner's licence, prospecting tools for recreational panning are restricted to the gold pan and shovel. Other methods - such as the dredge, sluice box, and rocker - undoubtedly process gravel more quickly than a pan and may be bought or built with portability in mind. Their use also involves more regulations.

3. REGULATIONS AND SAFETY

Like fishing, hunting, driving and just about everything else these days, prospecting comes with rules and regulations. Given that claim jumping might be the west's second oldest profession, that's no surprise! Also, the very nature of wandering around on unfamiliar ground means that common safety rules and respect for wildlife are relevant to the gold seeker's experience.

Much of the information that facilitates a goldpanning expedition comes from your local, friendly gold commissioner. Add a topographic map and some common sense and the rest falls into place.

VISITING THE GOLD COMMISSIONER'S OFFICE

A complete list of the Gold Commissioner offices throughout B.C. may be found in the appendix. This is a one-stop shop for both beginners and serious goldpanners. Mining records, reference maps and the relevant provincial government laws are available for your perusal. Fortunately, there's a handy-dandy booklet, Guide to Mineral Titles in British Columbia, that describes the various types of claim that may be staked as well as the requirements for maintaining a valid claim.

At the core of prospecting in all active regions of the province are the Placer Title Reference Maps (PTRM).

About maps

Maps are vital to a successful, safe expedition. Before setting off, buy the relevant topographic map and refer to the Placer Title Reference Map for your area of interest.

These maps identify documented placer claims and open territory. Because claims expire (or are cancelled) with time, map status is updated regularly.

Much like the federal Catalogue of Nautical Charts steers boaters toward the relevant chart, the province publishes an Index to PTRM to assist you in locating the maps that apply to your area of interest.

Because these maps show claim boundaries, they help you avoid trespassing. They also list active claim numbers - should you decide to seek permission to explore a registered claim. The details of ownership can be found in the Mining Records and you can do a title search for a nominal fee. Usually if you extend the courtesy of a query to a claim owner, you'll be granted access. (If you hit the motherlode, however, you may have a finder's fee to negotiate.)

You will also want to consider purchasing a Free Miner Certificate ($35 at time of writing). Just as a fishing licence allows you to bait a hook, this certificate is required to stake a claim or use more sophisticated tools. As mentioned, several acts are relevant to mineral tenure, but the guide booklet will get you started. Save the fine print for a rainy day.

Staking a Claim

For the more dedicated prospector, this is an exciting step. Staking a placer claim grants you the exclusive rights to surface minerals on a section of Crown land. Research and timing are essential ingredients to success; the regulations themselves are relatively straightforward.

It's worth noting that although placer claims come in various shapes and sizes, a normal, full - size claim measures 1,000 metres by 500 metres and is rectangular. Contrary to what you might expect, a placer claim is staked with only two posts. The area is marked by running a straight line along the length of the claim from an initial post to a final post. This line, called a location line, bisects the claim.

FULL SIZE CLAIM

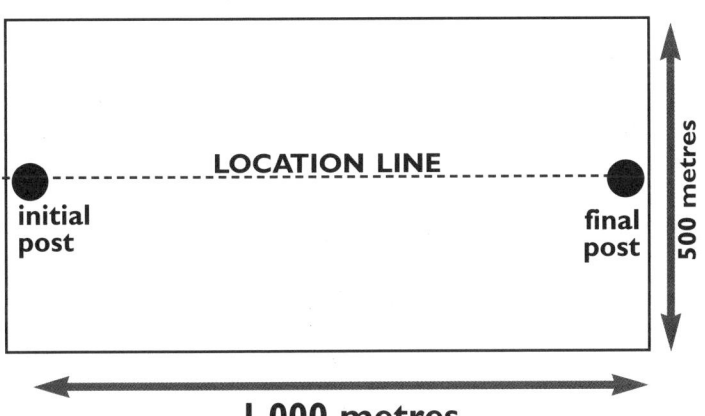

Following set procedures, making sure all the paperwork is done and paying the appropriate fee (currently $100 for placer rights) are all vital steps in securing a valid claim. Do your research and you should have no trouble. The GC will be able to supply the necessary forms and metal staking tags. Note that the Guide to Mineral Titles in B.C., while ideal for the recreational panner, is not a legal document. For the absolute gospel, review the Mineral Tenure Act and Regulations - which are also available from the GC or a B.C. Access Centre.

Maintaining Claims

**No placer claim lasts forever.
This is good news for the recreational panner
- previously staked areas are constantly coming open again.**

here's how:

Like other forms of ownership, a claim for surface mineral rights invokes certain responsibilities. The law sets out minimum annual requirements which must be met to maintain a valid claim. Otherwise, the claim is cancelled. It's as simple as that.

The annual requirements can be met in one of two ways:

1. Complete and record the necessary exploration and development work, or

2. Pay a fee (currently $500) in lieu of this work. The GC regularly reviews the mining records and neglected claims are cancelled on their anniversary date. Once cancelled anyone is eligible to restake and register their claim.

More Maps, Rights and Common Sense

Buy the relevant topographic map for your area of interest. These maps detail roads, rivers, creeks, elevations and land contours, and will help keep you on track. B.C. topographic maps are inexpensive and not too difficult to find, but you may have to go to a specialty map shop. Some bookstores also supply them. Hunt around in the larger urban centres or ask the GC office to recommend a nearby supplier.

BACKCOUNTRY RIGHTS & WILDLIFE

Other people's gold claims aren't the only property rights you'll encounter along the trail. Follow directions or seek permission when using logging or mining roads, ranchers' tracks or other private rights of way. When travelling in the backcountry, respect the environment. Close gates behind you and ensure that fences and cattleguards are secure.

Although bear or cougar attacks are relatively rare, stay alert. Most mishaps occur when people surprise an animal, so make plenty of noise. **Constant vigilance is essential if you are prospecting in grizzly country.**

LOGGING IN, LOGGING OUT

Before setting off let someone know where you are headed. If you don't know anyone in the area, notify the RCMP, the local fire department or district forestry office and always report back to your contact when your prospecting trip is done. Unless you are an experienced outdoors person, travelling alone isn't such a good idea.

A backcountry rescue can be a costly and time-consuming endeavour. If you do become lost or disoriented, though, it's best to stay in one place and wait to be found. Don't try to hike out, especially after dark. Light a fire, if it's permitted. Do what you can to keep warm and dry.

4. EQUIPMENT

When setting out on a prospecting adventure you should consult two separate check lists. The first one covers the basics - a list of musts for prospecting in the wilderness. The second list deals with optional equipment that will improve your chances of success.

THE BASICS

Clothing Loose fitting clothes, raingear and sturdy footwear are the hikers' second skin. Pack along extra socks too: your feet will likely get soaked.

Compass Essential: set the declination before striking out. Nearby metal (eg. a belt buckle, pocket change or vehicle) can affect compass readings so move away from any potentially magnetic source when you fix the compass setting.

Gold Pan A variety of sizes can be purchased or ordered through most hardware stores.

First Aid Kit Include bandaids, gauze, bandages and disinfectant. Depending on the season, consider adding insect repellant and bear spray.

Knife A sturdy folding or hunting knife is best.

Matches Pack a generous supply in a waterproof container.

Shovel A long-handled shovel is essential for digging gravel and sampling material below the surface.

Small Vial A film cannister or pill bottle is ideal for storing recovered gold. Fill the vial with water to prevent loss of fine gold in windy conditions.

OPTIONAL PROSPECTING GEAR

Altimeter

Available in both metric or imperial scales, an altimeter shows your elevation. To be accurate, the instrument must be set at a known point above sea level. An altimeter is affected by barometric conditions so it has to be reset daily. Topographic maps have contours to show elevation - at known locations you can reset your altimeter to these.

Hammer

Carried on your belt, a geologist's hammer (hammer face on one end, pick on the other) is great for working creek banks and breaking rocks.

Magnifying Glass

A 10-power glass will help spot fragments of gold. A folding, pocket model is ideal.

Magnet, Tweezers

A magnet will remove black sand from gold concentrate; wrap it in plastic for easy cleaning. Tweezers handle coarse gold that's too small to be grasped with fingers.

Plastic Sheet

A large, plastic sheet can be useful for prospecting high banks and will double as a rain cover or tarp.

Pry Bar

A small pry bar or chisel will widen crevices and cracks in bedrock. The chisel can be worked into deeper areas to recover trapped gold.

Suction Gun, Brushes

A syringe-type suction gun with a long tube can be used to draw material out of hard-to-reach places. For wider cracks, a small brush with stiff bristles can be also used. Some prospectors pack along narrow paint brushes for this purpose. A kitchen knife or steel spoon will also prove handy.

5. PROSPECTING TALK - LEARNING THE LINGO

Like any field, prospecting has its own special terms. Some of these you'll encounter in the text.

Auriferous: Gold-bearing.

Bars: Each year more gold is deposited in these sand or gravel areas. There have been some spectacular gold discoveries in river bars.

Bedrock: The rock bottom of a stream or river.

False Bedrock: An area of heavy clay or cemented gravel overlying real bedrock.

Benches: Areas of flat or slightly inclined ground lying immediately above creeks and rivers. The banks can yield excellent returns.

Black Sand: Iron concentrate, usually composed of hemetite or magnetite.

Coarse Gold & Fine Gold: Nuggets, usually discovered close to their original source. Fine gold has usually travelled a greater distance and is more rounded and flattened by the action of water.

Crevice or Crack: A fissure in rock where gold traps and accumulates.

Fineness: A term to determine the purity of gold - 1,000 fine being the top of the scale. Gold's fineness in most streams ranges from the low 700s to the high 900s.

Float: The material washed down from high banks during runoff.

Fool's Gold: Iron pyrites or mica.

Fraction : A mining claim that's less than full size. A fraction usually occurs when an area is staked between two or more existing claims.

Gradient: The descending slope of a river or creek, denoting its elevation.

Grizzly: A heavy screen placed on top of a framework to prevent rocks from entering sluice boxes. (If someone yells grizzly and hightails it, though, it might mean something else!)

Hydraulic Mining: Water forced under high pressure to wash hillsides and creek banks into sluice boxes. This method of mining is no longer allowed.

Lode or Lode Vein: Gold veins in solid rock are the source of most placer gold found in streams.

Moss: Another potential source of trapped gold; break it up and pan it.

Nugget: A lump of solid gold that can be minuscule or weigh many ounces.

Oil: Fish oils sometimes coat gold, especially in large spawning rivers such as the Fraser. Gold coated in oil sometimes floats.

Overburden: The non gold-bearing gravel, rocks and clay overlying gold which has settled closer to bedrock.

Quartz: Most gold deposits are veins found in quartz, which can be colourless, white, rose coloured, brown-stained, or green. Also called wire gold.

Rocker Box: A sieve mounted on a sloping rocker base which has riffles or canvas collection blankets to trap gold.

Sluice Box: An open-ended trough with riffles on the floor, positioned in the creek so that water washes over the river gravels shovelled into the box.

Sniping: Cleaning old mine dumps and tailings, as well as cracks and crevices in bedrock.

Tailings: Dumps, mounds and ponds at old mine sites. Tailings have been known to produce nuggets of considerable size.

Troy Weight: Also known as apothecary weight, this system, commonly used by assayers, is based on the pound containing 5760 grains, or 12 ounces of 480 grains each. By contrast, the more commonly used avoirdupois or U.S. customary system of weight is based on 7000 grains, or 16 ounces of 437.5 grains each.
In troy weight, the system used for precious metals and gems:
24 grains = one pennyweight
20 pennyweights = one troy ounce
12 troy ounces = one troy pound
1 troy pound = 0.8232 U.S. customary pounds, or 373.24 grams.

**Miner washing gravel
with a grizzly**

THE CARIBOO

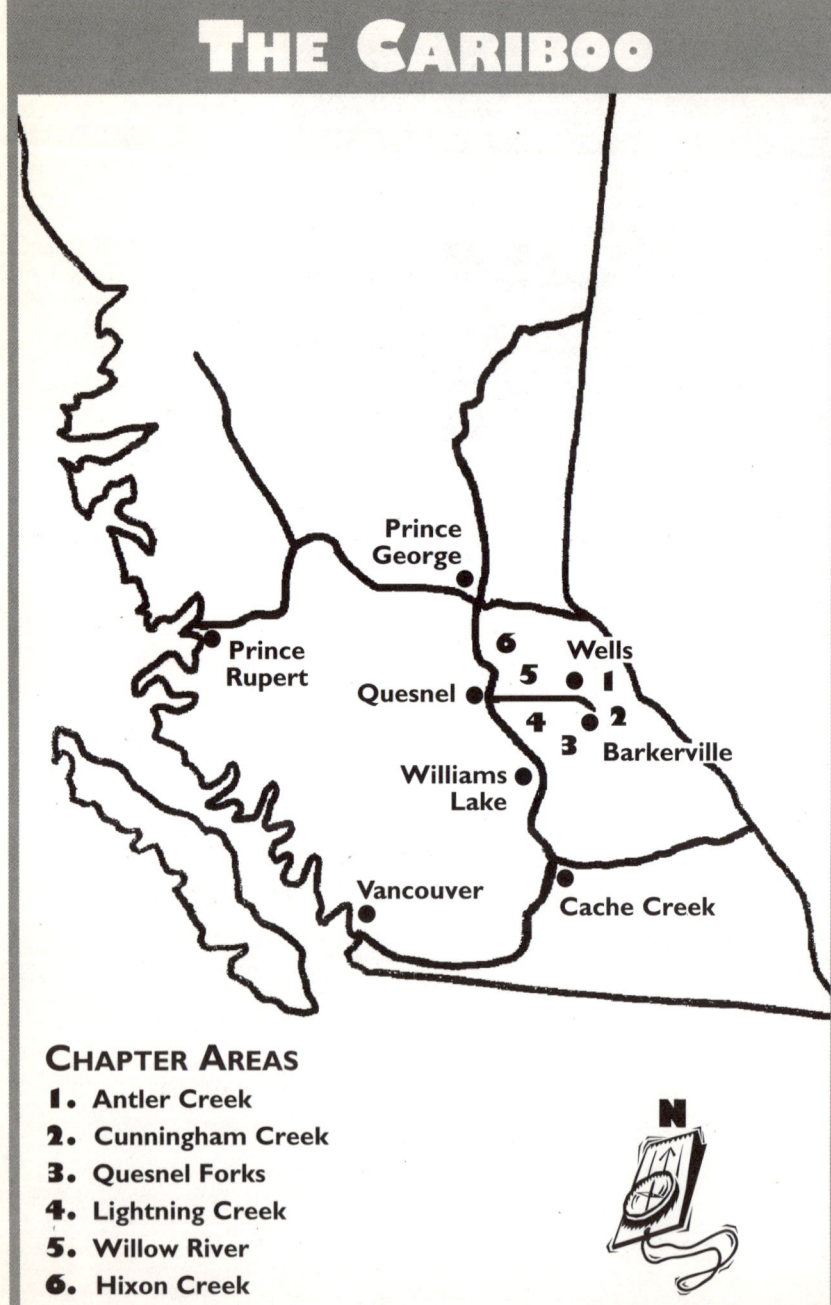

Prince George

Prince Rupert

Quesnel

Wells

6

5

4

1

2

3

Barkerville

Williams Lake

Vancouver

Cache Creek

N

CHAPTER AREAS

1. **Antler Creek**
2. **Cunningham Creek**
3. **Quesnel Forks**
4. **Lightning Creek**
5. **Willow River**
6. **Hixon Creek**

$\mathcal{T}$HE CARIBOO'S GOLD-BEARING CREEKS

The gold history of the Cariboo is legendary and has been well documented. Over the years, many thousands have prospected and mined the slopes and valleys of this rich country. In spite of all the attention, though, the gold-bearing creeks of the Cariboo have never been worked to their full potential.

Until relatively recent times, it wasn't easy to track the Cariboo's gold secrets; the lack of roads and easy transportation made the region inaccessible to all but the most determined and hardy prospectors travelling on foot. These adventurers courted fortune with little more than the shirts on their backs and the most basic of tools.

The original channels that were worked, and there are some that are still untouched, were frequently exploited only for their most accessible gold. As soon as widening or deepening channels became too difficult to mine efficiently with the primitive tools available, they were abandoned in favour of more promising locations.

Along with the physical hardships, miners have also had to contend with the host of scoundrels that gold seems to attract. As a commercial venture, the lure of gold can still prove a risky game. The unwary and uninitiated who have dabbled in placer operations on the Vancouver Stock Exchange can attest to that.

For today's recreational panner, however, the possibilities are abundant. Roads and trails continue to open up the backcountry, and with this book as your guide, you'll find the creeks, streams and channels of the Cariboo are by no means played out.

Panning for gold is a fascinating, enjoyable and rewarding pursuit. The magic dust, alone or bound up with other minerals and metals, is a beautiful and mysterious substance that displays a different fineness and texture in every creek of the province.

What follows is a selection of panning gems, derived from the personal diaries and recollections of a dedicated prospector. They are confidently recommended on the basis of extensive experience prospecting in the field, and a thorough review of the relevant mining history.

This volume focuses specifically on golden opportunities that are relatively easy to access.

For ease of reading, two conventions have been used throughout:

- all distances are estimates;
- directions assume you are facing upstream.

Each location is also coded:

▲ - means a short step away,
▲▲ - indicates a longer expedition.

The chapter maps are designed to provide a reference for specific creek locations. Due to size limitations, however, it is impossible to include all the creeks mentioned on these maps. It is recommended that you acquire more detailed maps when visiting specific sites.

Good luck wherever these leads take you, and remember to have fun!

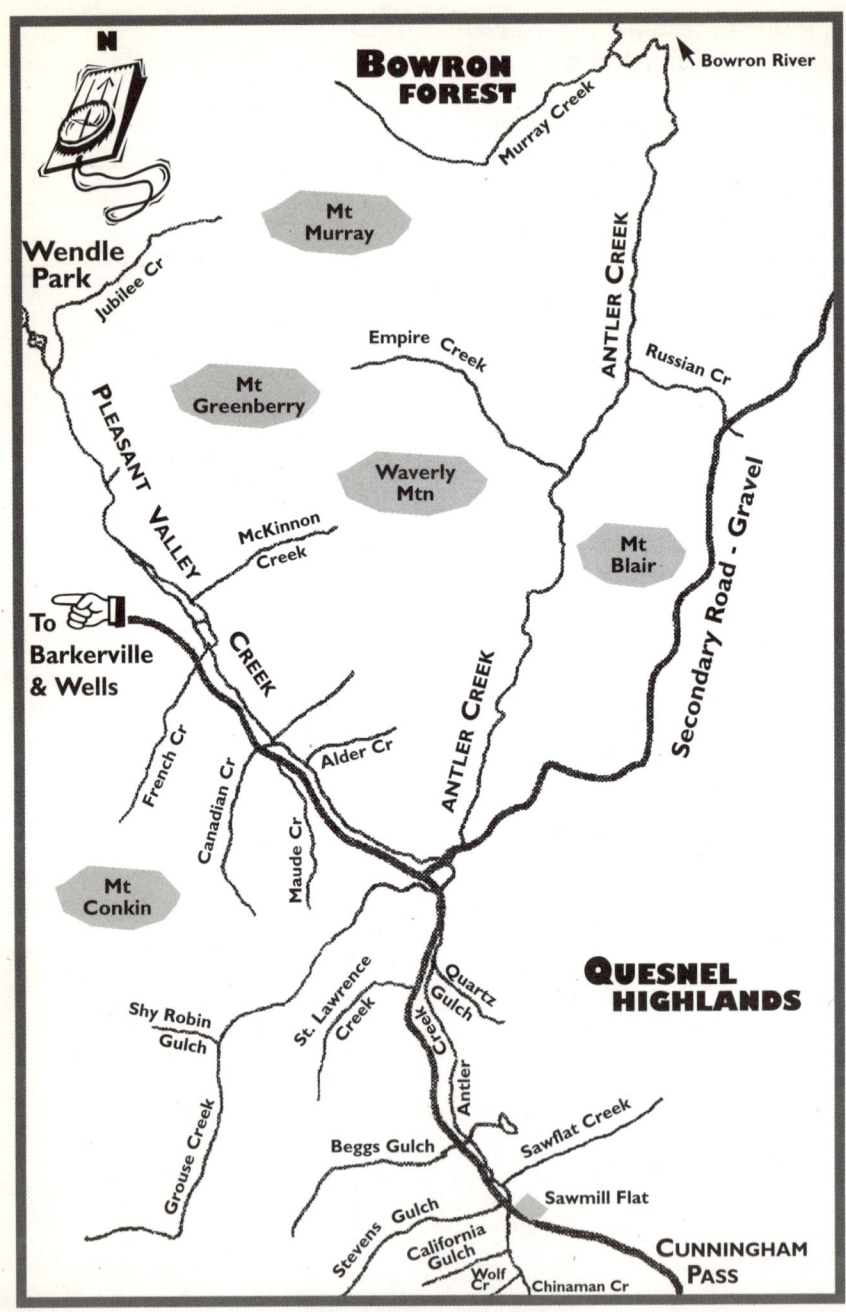

ANTLER CREEK

*I*f your pickup's worthy of the roads out in the country,
Head out to the Cariboo and stake yourself a claim.
Fill your knapsack up with food
And head out while you're in the mood,
See if you can beat old 'Lady Luck' at her game.

- from *Go for the Gold*, by Jean Ann Robitaille

*A*NTLER CREEK

The headwaters of Antler Creek, located between Mount Burdett and Nugget Mountain, are accessible by road from the town of Wells, east of Quesnel on Highway 26. A main road, secondary roads and hiking trails cross Antler Creek in several places before it flows into the Bowron River below Bowron Lakes. Although access is now relatively easy, many tributaries and promising sections of Antler Creek itself have still not been worked. It's a little-known fact that many of the old, buried channels in the Antler Creek area of the Cariboo have never been mined for their gold.

ANTLER / CUNNINGHAM PASS ▲▲

In 1882, the Yellow Lion Company was operating in the Antler / Cunningham Creek Pass and found gold on bedrock within a few feet of the surface. Yellow Lion miners traced this find into an old channel in the pass and continued along for a quarter of a mile, working by hand. The claim paid well. In spite of this success, the company never prospected the next three or four miles of Antler Creek.

The Ne'er Do Well Dump, Grouse Creek, 1868

The discovery of gold on the Quesnel River in 1859 led to major finds on Antler Creek the following year. In 1861, with gold valued at $16 an ounce, Antler Creek was yielding as much as $10,000 a day - $250,000 at today's gold prices.

BLIND CHANNEL ▲

Try prospecting 2.5 miles upstream from Antler Creek's mouth on the Bowron River. In 1896, a blind channel was discovered here in the left bank of the creek. This channel was prospected, but never mined.

SAWMILL FLAT ▲

An extensive, ancient high channel is located in this area. Past mining evidence suggests that the old channel which enriched Antler leaves the creek's present course on upper Antler Creek. From there it runs through the low pass into Cunningham Creek's valley. In 1901, at the junction of Wolf and Antler creeks, two half-mile leases ran parallel to Antler south towards Sawmill Flat and were producing well. An old channel here, high above creek level, has never been mined. Mining records in 1902 also show encouraging results at the lower end of Sawmill Flat.

By 1878, several companies had staked claims along the benches of Antler Creek about 10 miles downstream from its source. These benches are quite extensive and the companies that worked them enjoyed good rospects for a number of years.

CHINAMAN CREEK ▲▲

The whole creek is worth prospecting. Chinaman Creek's headwaters lie close to Cunningham Pass. It's a tributary of Wolf Creek, which flows into Antler Creek from the west just below what used to be the Nason claim. There, a quarter of a mile and some 200 feet above the Antler Creek valley, Chinaman Creek's old channel was discovered at the turn of the century and it proved to be one of the most rewarding properties in the area. The channel is about 200 feet wide and 70 to 80 feet high. The whole area still looked extremely promising prospecting in the 1970s.

Construction of the Canadian Pacific Railroad and the lure of a wage drew many miners away from the gold fields in 1885. In 1887, only the Nason Company was working on Antler Creek.

Pleasant Valley Creek was discovered in 1896 and yielded excellent results.

RUSSIAN CREEK ▲▲

A bench more than a mile long and ranging from 500 to 1,000 feet wide was discovered at the junction of Russian Creek and Antler Creek in 1906. The miners who worked this bench channel found that gold values were increasing with depth, but they never did reach bedrock - where the largest deposits accumulate.

In 1919 very high gold values were found in the Antler Creek bed 12 miles north of Barkerville. A dredge was proposed, but in the end was never built.

GROUSE CREEK ▲

The east bank of Antler Creek, half a mile upstream from the mouth of Grouse Creek, is another promising area. A small amount of work here yielded sparkling returns in 1908.

ALDER CREEK ▲

New discoveries were made on upper Antler Creek in 1926. Three-quarters of a mile from the mouth of Pleasant Valley Creek, prospect Alder Creek's right bank. An old channel was discovered here apparently crossing Alder Creek in a more or less north-to-south direction. On the east side, this old channel is about 120 feet wide and runs south to a point opposite Wolf Creek

The Nason Company Mine, Antler Creek, 1902

- where it may once more cross Antler Creek. Three-quarters of a mile west of this area, on the other side of Alder Creek, the old channel is again exposed. In 1927, water was piped here from China Creek, Wolf Creek and Stevens Gulch and the channel produced well.

The 1921 mining records show that even on well known creeks there was still plenty of virgin ground. For the next three years, keystone drilling, a method of core sampling, was carried out on Antler Creek and produced great results.

CALIFORNIA GULCH ▲

Coarse gold was found on the rim of California Gulch in 1929, but a pre-glacial channel, which starts in the vicinity of Moloney Flat, has never been worked. This channel parallels Antler Creek's present course, joining it between the road and the lower end of China Creek. Three-quarters of a mile below Sawmill Flat part of this channel is exposed, but roughly 2.5 miles of the channel is still intact - it has never been disturbed. This is located in one of the richest areas of the Cariboo!

In 1932, a private company managed to work the exposed section of the channel where it crosses Antler Creek diagonally, and it paid extremely well. About 260 feet above creek level the bedrock is exposed and the channel lies right next to it. In the early 1970s, the creek portion was staked at this location, but the old channel itself was not. Exploratory panning here produced good results in the bottom 10 feet of the channel, just above creek level.

In 1950, two small-scale operations were on Antler Creek; upper Antler Creek had just one lease in progress. Lack of machinery limited the size of these operations.

SAWFLAT CREEK ▲

A significant discovery was made near the junction of Sawflat Creek and Antler Creek in 1948 when bedrock samples from keystone drilling (core sampling) were examined.

From the 1960s to the present not much work has been done on Antler Creek. The biggest obstacle in recent times has been the lack of financing for placer operations.

The samples provided conclusive proof that in pre-glacial times Antler Creek, upstream from Sawmill Flat, flowed down the valley of Sawflat Creek and into the Swift River. Like other promising areas in the Antler Creek region, this old channel was prospected, but has never been mined.

The boys on the boardwalk of Barkerville, early 1900s

HAZARDS

"I find in the winter time you've gotta be a lot more careful, especially if you're in avalanche country. Because I got caught in one once with a snowmobile.

The slide came down ... I didn't see it coming at first because of the noise of the snowmobile - you know, you're not looking up like that at the side of the mountain - and all of a sudden it hit me.

Luckily ... I went down ahead of the slide and the snowmobile kept going ... I got tossed off it anyway, one way or the other, and I landed in the snow and I kept rolling and trying to run.

The chap I was with was ahead of me, maybe by 50 or 60 yards, and he didn't get hit with it, but I did and the snowmobile went one way and I went the other.

I left that snowmobile right there - it was an older snowmobile. It was under the snow somewhere. It quit so I couldn't hear it running and I thought: 'To hell with it, let's get outta here before something else comes down.'

I never saw that snowmobile again. As far as I know it's still there to this day!"

- Jim Lewis

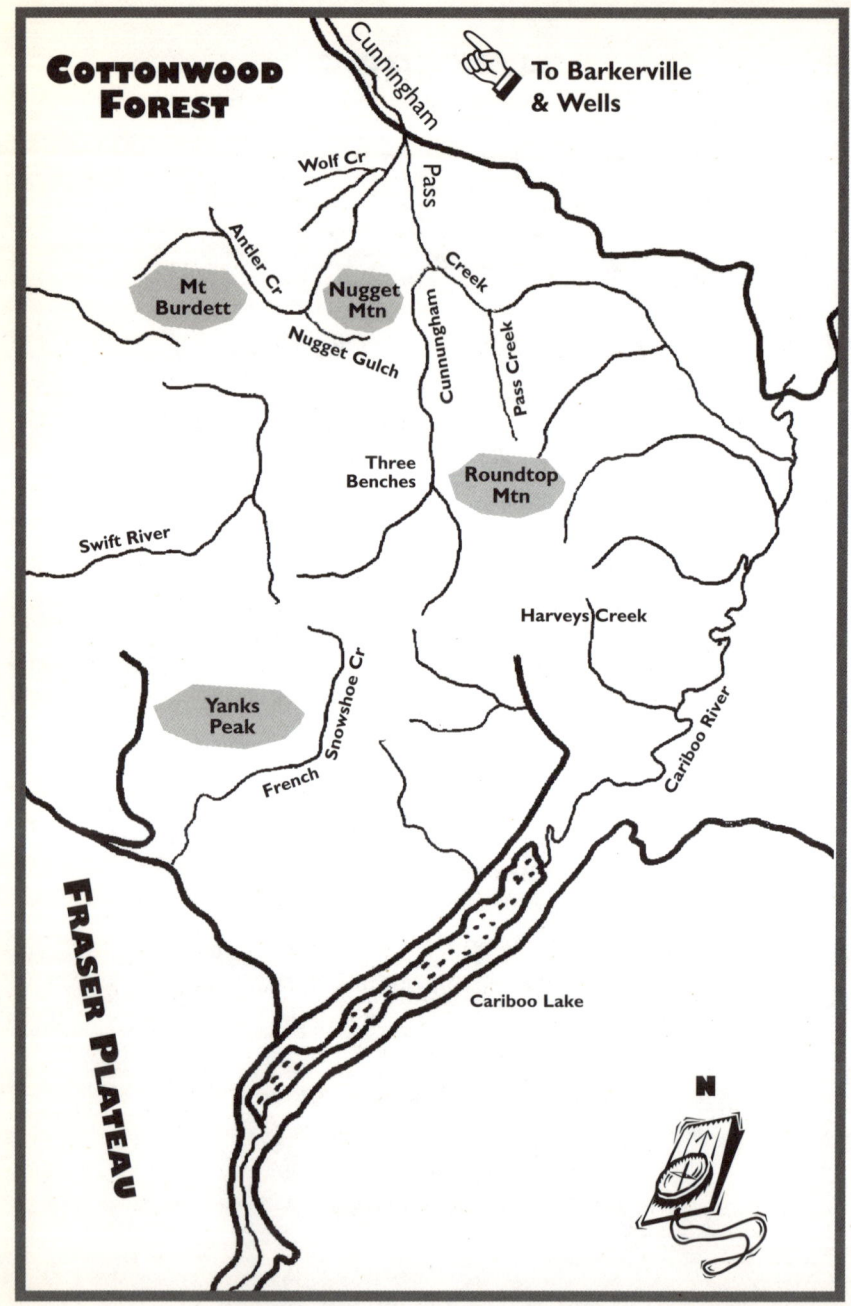

CUNNINGHAM CREEK

$\mathcal{B}$ut if you get discouraged 'cause the gold's not free and easy
Sit back a minute and take a look around.
'Cause if you felt the mornin' sun
And if you saw the wild deer run,
You've struck gold when you've found
All the gold's not in the ground.

- from Go for the Gold, by Jean Ann Robitaille

$\mathcal{C}$UNNINGHAM CREEK

The headwaters of Cunningham Creek are located between Yanks Peak and Roundtop Mountain, clearly defined on the topographical map for the top of Quesnel Lake. A review of Cunningham Creek's gold history shows that in spite of its productivity over the years relatively little work has been done. There are many buried channels in the area. Given the hardships borne by the first prospectors - lack of transport, lack of cabins in the wilderness, lack of financing, lack of water to adequately work claims, lack of machinery - it's not surprising that these buried channels have not been mined out.

CUNNINGHAM CREEK PASS ▲

In 1888, 2.5 miles from Cunningham Creek's mouth, paying gold was found 350 feet up the hillside in the low pass leading to Cunningham Creek. Hand methods were producing 1.5 ounces of gold a day for each man working the area at this time!

Travelling through Cunningham Pass from Antler Creek, the flats are underlain

The most famous gold pan in the Cariboo belonged to William Cameron, of 1858 gold rush fame

The Tregillus family, (below) gathered for a portrait in front of their Cunningham Creek cabin in 1929, worked one of the few leases in the area at the time

with gravel deposits. In the early 1970s I made a good living working through there with my old prospecting partner, Pat Harvey. During the dry summer months, using a portable 2.5 inch dredge, we were retrieving as much as fourteen ounces of gold a day working no more than three feet below the surface. In a few places, the pass had previously been prospected, but most of the gravel deposits had never been touched.

CUNNINGHAM CHANNEL ▲▲

There is little doubt that an old channel leads through the low pass into Cunningham Creek's valley. Just below the pass, on the west branch, the creek takes a sharp turn to the east. Here, there's a 40-foot bank of gravel which proved quite lucrative at the turn of the century. Paying quantities of gold were also found above the west branch, on high benches. The old channel meets creek level four miles upstream from the west branch. Diggings at this location are relatively shallow; prospecting in more recent times produced results at depths of four to six feet.

OLD STREAM BED ▲

In 1886, Cunningham Creek's old stream bed was discovered 250 feet above creek level and it produced well. Quartz veins containing gold were also found two miles west of Roundtop Mountain. A logging road parallels the creek in this location.

In 1907, a dam constructed the previous year burst, halting all work on Cunningham Creek.

CUNNINGHAM BAR ▲

At the turn of the century, in an area known as Cunningham Bar (directly east of Nugget Mountain), rich gold deposits were discovered on bedrock. Banks 150 feet high were producing well. Progress was limited to 50 feet a year, however, due to the primitive methods and lack of water for hydraulic mining. Prospecting here in the 1970s produced sparkling results in the gravel banks - starting at depths of about four feet.

There is little record of further mining activity until 1921, when a small scale operation was underway on one lease. There was still plenty of virgin ground on many of the well known placer creeks in the area.

PASS CREEK ▲

An old channel lies in the south side of Cunningham Creek half a mile downstream from its confluence with Pass Creek. Records for 1904 show the gold recovered ranged from fine dust to nuggets weighing one-third of an ounce. Diggings are relatively shallow at this location.

NUGGET GULCH ▲

In 1905, a pre-glacial channel was discovered in Nugget Gulch coming in from the head of Cunningham Creek. It's very likely the source of the gold found there.

Drilling in 1924 proved up some three million yards of gravel on Cunningham Pass Creek with good gold values. This was drilled only, never mined.

OLD CHANNEL ▲▲

In 1929, Cunningham's old channel was discovered running parallel and west of the existing creek. Both rims of the old channel were well defined, but prospectors failed to trace it to the location where it intersects the present creek. It's believed the channel extends some two miles downstream from Cunningham's mouth. This area would be well worth exploring.

THREE BENCHES ▲▲

Drilling in 1930 and 1931 yielded unusually good results in three benches located two miles southwest of Roundtop Mountain, at the confluence of Cunningham and an unnamed creek. On the map you'll see that the road parallels the creek at this location. The first bench was located just a few feet above the creek, the second one 140 feet above creek level, and the third bench is 20 feet higher again. These benches, which appear to be undisturbed by glacial action, are at least 130 feet wide and extend away from the creek for some distance. The benches yielded noteworthy results in 1932 and, a year later, there was new evidence that the old channel tends to run into the right bank of the creek here.

In 1941, hydraulic miners tested a gold-saving device designed to recover gold from black sand. The idea was later tried on the Fraser River south of Quesnel.

The recorded mining history indicates only limited work on Cunningham Creek since the Second World War.

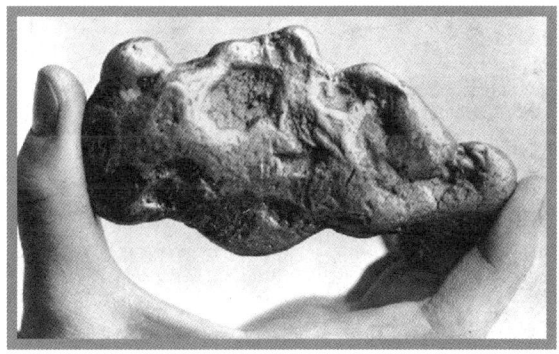

An egg-size nugget found in the Cariboo

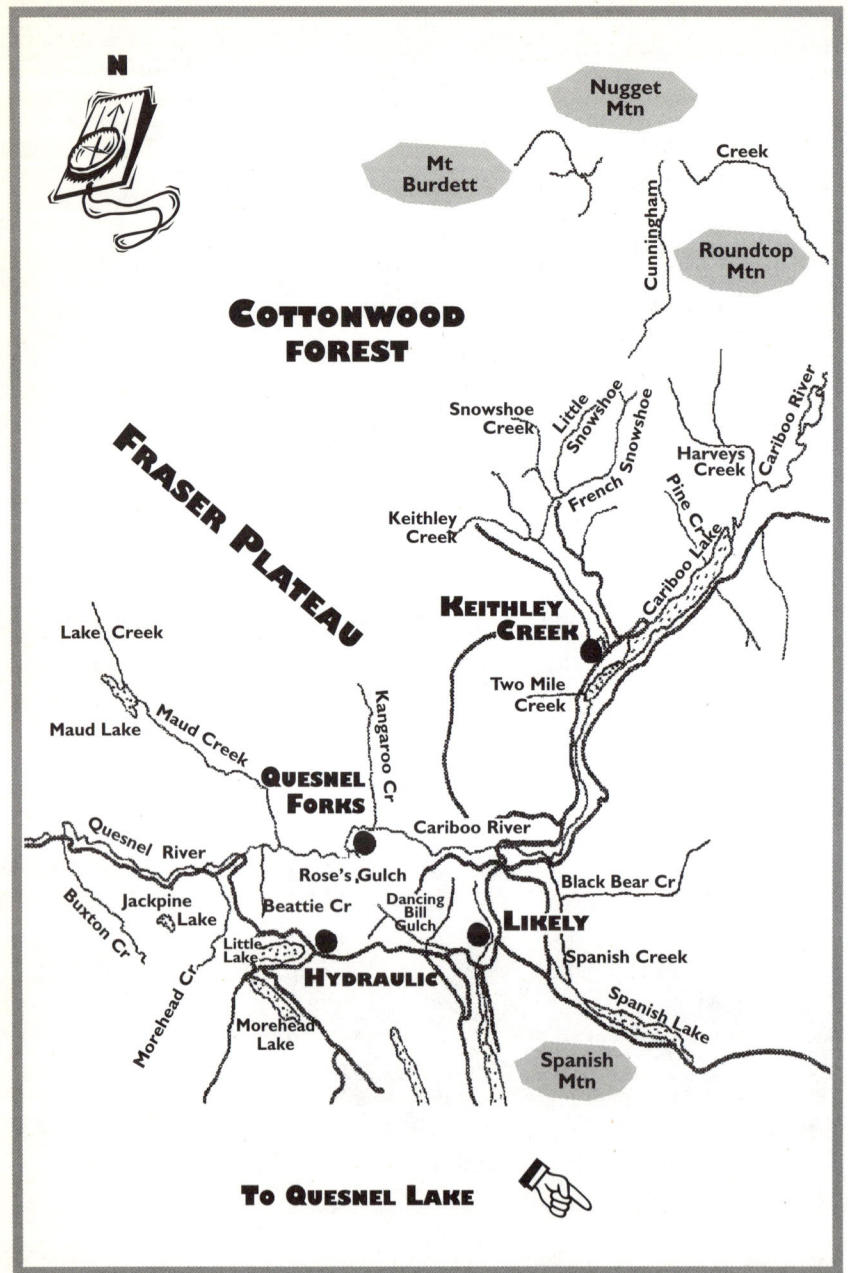

N

Nugget Mtn

Mt Burdett

Creek

Cunningham

Roundtop Mtn

COTTONWOOD FOREST

Snowshoe Creek

Little Snowshoe

French Snowshoe

Harveys Creek

Cariboo River

FRASER PLATEAU

Keithley Creek

Pine Lake

KEITHLEY CREEK

Cariboo Lake

Lake Creek

Two Mile Creek

Maud Creek

Maud Lake

Kangaroo Cr

QUESNEL FORKS

Cariboo River

Quesnel River

Rose's Gulch

Black Bear Cr

Buxton Cr

Jackpine Lake

Beattie Cr

Dancing Bill Gulch

LIKELY

Spanish Creek

Little Lake

HYDRAULIC

Spanish Lake

Morehead Cr

Morehead Lake

Spanish Mtn

TO QUESNEL LAKE

QUESNEL FORKS

They say she's richer than the '49, go there to sink your pegs
And all you need to get there is a strong pair of legs
They are finding placer gold and nuggets as big as eggs
In the creeks that send their waters to the Fraser
                                                  - from The Mighty Fraser, by Roger Koe

CHAPTER 3

UESNEL FORKS

Just downstream from the town of Likely, for a time in the later part of the 19th century the Quesnel Forks area was almost exclusively a Chinese settlement. Numerous small creeks in the area are rich in possibility and rich bars have been found all along the south fork from Quesnel Lake. Roads now lead all along the river and the valley is wide. The surrounding hillsides still contain extensive gold-bearing gravel deposits, 700 feet above the valley floor.

Along with the specific sites mentioned, the following creeks are also known to be gold-bearing: Red Gulch Creek, Snowshoe Creek, French Snowshoe Creek, Goose Creek, Duck Creek, Spanish Creek, and Cedar Creek.

20 MILE CREEK ▲▲ (not on map)

In 1883, gold was discovered at 20 Mile Creek (west of Buxton Creek towards the Fraser River). At the time there were neither roads nor trails into the area and river travel was a hazardous undertaking due to the river's swift current and many boulders. Prospectors still managed to take excellent returns a few years later. This time the action was on the south side of the Quesnel River, just below 20 Mile Creek. By 1894 there were numerous leases in the area, but little or no work was being performed on them.

High Benches ▲

In 1875, miners were producing a third of an ounce a day panning on large benches 50 to 200 feet above the river.

Six miles above Quesnel Forks on the Cariboo River, cuts in the north side high benches yielded excellent results in 1896. My own prospecting more than 80 years later showed the river banks at this location are still producing.

Bullion Mine ▲

The Bullion Mine managed by J.B. Hobson came into production in 1897 on an old buried channel on the south fork of the Quesnel River. This channel (see map, page 54) still exists. The Bullion property originally comprised eight leases (466 acres) and extended nearly two miles along the channel's west side. The bars and low benches of the river here have been enriched by a series of criss-crossing channels, some of which have never been mined. Spring 1997 mining records revealed some portions of this channel were open for staking!

Guards show off gold specimens and guns, Bullion Mine 1902

DANCING BILL GULCH ▲

(Like Williams Creek, this one's named after Bill Williams, who'd do a little dance every time he found good gold in the pan!)

A large gravel deposit follows this small stream to the river, crossing an ancient river channel and exposing gold-bearing gravels. The old channel can be traced for 1.5 miles in each direction - nearly paralleling the south fork of the Quesnel River. A rocky ridge, once known as French Bar Bluff, separates the channel and the river. Upstream from its intersection with Dancing Bill Gulch, this channel turns abruptly into a more recent channel. After crossing this newer channel, the gulch follows it on its turn into the river. Recent dredging and sluicing produced excellent results through here.

In 1878, very little placer mining was done - most miners were caught up in a mini gold rush examining hard rock possibilities in the area.

High water meant very little mining was done in 1882, but a dry year in 1883 allowed previously unworked benches to be mined.

BLACK JACK GULCH ▲

Half a mile southeast of Dancing Bill Gulch, the channel just mentioned crosses Black Jack Gulch. However, this gulch didn't cut deep enough through the rim rock to expose the channel. Gravels I tested here proved rich in gold, ranging from fine dust to nuggets weighing a quarter of an ounce. Some of the gold was flattened and well worn; other pieces were in milky white quartz and resembled the vein gold found all over the Cariboo.

In 1885, many Chinese were working productive claims; 1886 was another dry season and the bed of the river at Quesnel Forks was producing well.

CHANNEL DEPRESSION ▲▲

To the southeast, this same channel runs in a well-defined depression for about half a mile until it meets another

South Fork of Quesnel River - 1927

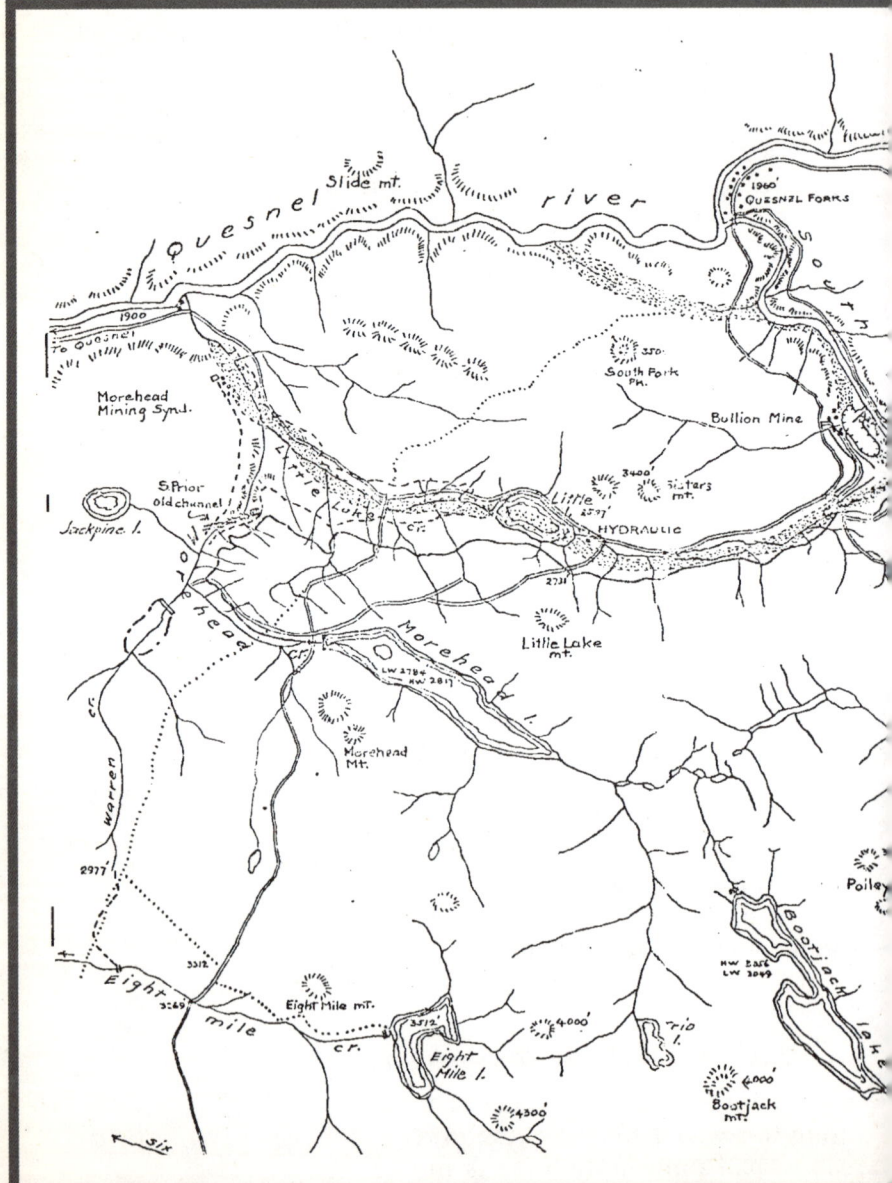

This map, from Jim Lewis' mining library, typifies maps that accompanied mining engineering reports dating back to the 1880's.

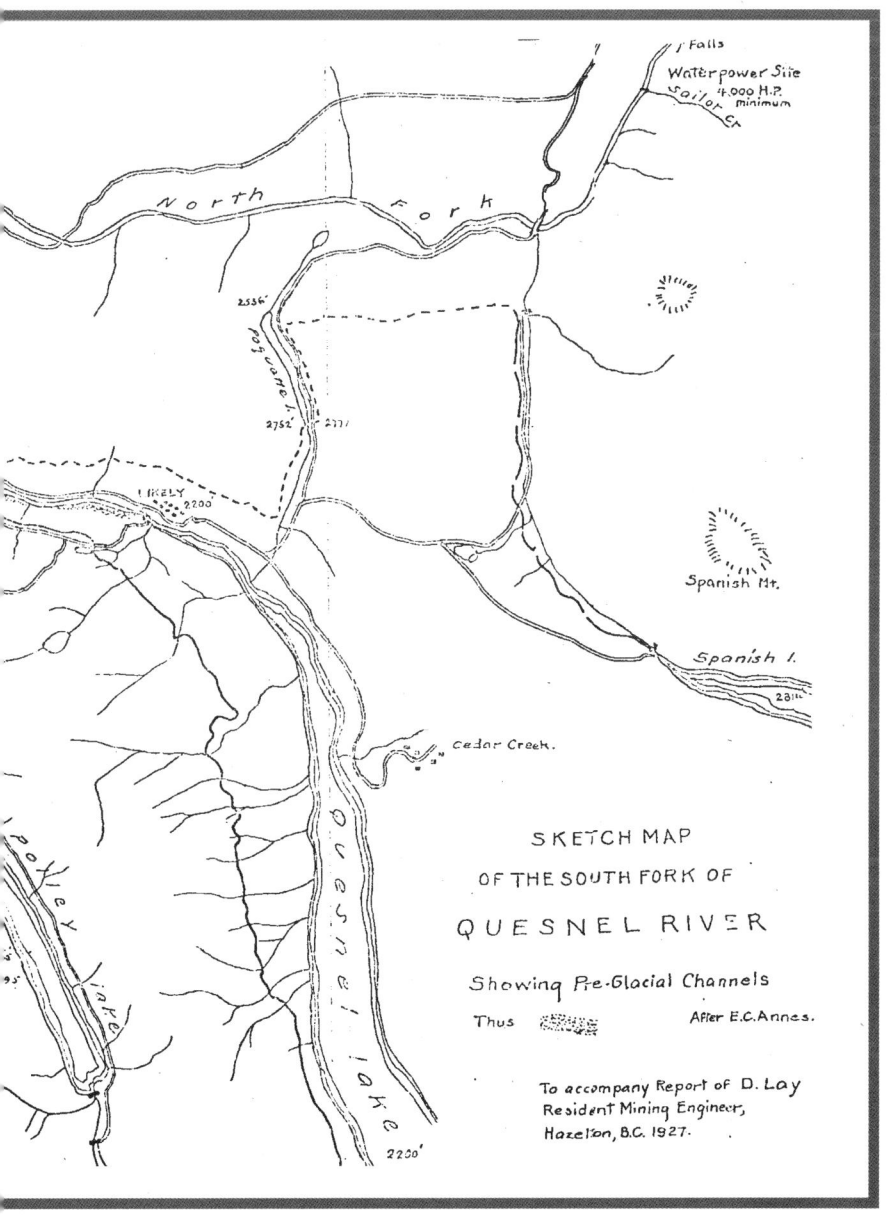

SKETCH MAP
OF THE SOUTH FORK OF
QUESNEL RIVER

Showing Pre-Glacial Channels

Thus ░░░░   After E.C.Annes.

To accompany Report of D. Lay
Resident Mining Engineer,
Hazelton, B.C. 1927.

In 1893, reports indicate the north fork of the Quesnel River (Cariboo River) was producing well.

depression occupied by Little Lake. Little Lake Creek runs west from here, entering Morehead Creek two miles upstream from the Quesnel River. The course of the old channel should be thoroughly prospected through this area.

ROSE'S GULCH ▲

On the other side of the south fork, along Rose's Gulch, another old channel contains a significant deposit of gold-bearing gravel. Although it was discovered 100 years ago, recent explorations proved this channel's still producing.

BEATTIE CREEK ▲

East of Beattie Creek, on the south bank of the main Quesnel River, a large gravel deposit sits in what's believed to be part of the ancient river system. Gold bearing, this ancient channel runs through a low, valley-like depression a mile long and ranging from 2,000 to 3,000 feet wide.

In 1896, Kangaroo Creek was found to be rich from the surface down.

I've traced this channel and found it still intact, with paying values.

MOREHEAD CREEK ▲

Near its mouth (six miles downstream from Quesnel Forks) there's evidence of another huge deposit of gold-bearing gravel in an old river channel. Three-quarters of a mile upstream from the confluence, gently sloping ground rises on to a steeper line of hills - 400 to 500 feet above the creek. Here, in a depression, Morehead Creek has cut a narrow gulch

Around 1901, workings on Spanish Creek recovered as much as 12 ounces of gold every 10 feet.

and there's a deposit of stratified gravel and gravelly clay that's 2,000 feet wide and 100 to 150 feet deep. Morehead Creek cut through the bedrock in places and, not far from the western rim, also cut through this stratefied deposit leaving benches of gravel. Without equipment to mine the higher ground, earlier work here was confined to the rich gravels in the creek bed. My own explorations also revealed evidence of other gold-bearing channels in the area.

BEAVER RIVER ▲ (not on map)

At the mouth of Beaver River, west of 20 Mile Creek, mining records note paying ground on high benches on the south side of the Quesnel River. This ground was never fully worked, though, due to labour shortages at the time of discovery. It was consequently still producing paying quantities not long ago.

**Abandoned
buildings at
Quesnel Forks,
1979**

Overlooking the Bullion Mine, Quesnel River 1890s

1880s Chinese settlement nestles on the bank of Keithley Creek

LAKE CREEK ▲▲

A large creek that flows on a steep grade into the head of Maud Lake (previously known as Four Mile Lake). Coarse gold has been found here at a point where the creek's ancient channel is indicated in the left bank. The ground was also worked just downstream from where Lake Creek emerges from a rocky gorge. An ancient buried channel segment, plainly indicated here in the right bank, offers good possibilities.

MAUD CREEK (FOUR MILE CREEK) ▲▲

Maud Creek meets the Quesnel River four miles downstream from the Forks. About a mile upstream from here, I've located gold deposits in gravel. These deposits likely emanated from an ancient channel believed to run along the north side of the Quesnel River.

In 1912, with the war approaching, there were less miners around. Work was proceeding on Spanish Bars and Marten Creeks, however, and Keithley, Goose and Four Mile (Maud) creeks were being prospected.

NORTH BANK ▲

Four miles east of Quesnel Forks, along the north bank of the Cariboo River, a large and thick gravel deposit is exposed by the river. There's good paying gold here as well.

KANGAROO CREEK ▲▲

Flows into the south fork of the Quesnel River half a mile upstream from the Forks, on the north side. The topography suggests that a pre-glacial channel of considerable length lies buried in the left bank of this

Very little mining was done in 1913, but two new areas are mentioned: Honey Creek and Half Mile Gulch.

creek. My own prospecting unearthed a remnant of this channel.

POQUETTE CREEK ▲ (not on map)

Poquette Creek cuts through extensive gravel deposits a few hundred feet upstream from its mouth on Quesnel Lake. Likely Gulch, a small tributary of Poquette Creek, flows in from the east a mile upstream and is also worth checking out. A small flume and gravel washing boxes produced paying values here at one time. Even so, relatively little work was done.

BLACK BEAR CREEK ▲

Testing in the valley of this stream, which lies east of Poquette Creek, indicated appreciable gold values. On the north side of Black Bear Creek, earlier sluicing was successful on what appears to be the opening of an old, high channel running into the creek's valley. Old channels are also in evidence along Spanish Creek, which joins Black Bear Creek about a mile south of the Cariboo River.

From the mining records in the early 1900s: "The country drained by Keithley Creek and its tributaries has not gained the attention of the capitalists, but the gravels in the section are rich and should attract the attention of large operators when transportation becomes easier."

KEITHLEY CREEK ▲

Flows into Cariboo Lake from the northwest. A channel discovered in the hillside 600 feet above the old creek bed at the turn of the century still exits today. In the 1930s, gold values in gravel were discovered on the left bank of the creek half a mile upstream from the bridge on the Likely Road. The gravels overlie a rock bed flanking the creek.

LITTLE SNOWSHOE CREEK ▲

The upper and main tributary of Keithley Creek, which it enters from the north east. I've prospected extensively through Little Snowshoe Creek and it's especially rewarding two miles upstream from its mouth. There, on the left side, an old high channel appears to cut into the hill in a southerly direction.

UPPER PINE CREEK ▲▲

Gold-bearing gravels were first discovered here in 1935 and the whole area is still excellent, especially the high banks. The creek flows southeast in a narrow V-shaped valley, eventually turning abruptly northeast half a mile from Cariboo Lake. Here it enters a short canyon, emerges on to bedrock, then flows on for some distance before turning southeast once more to meet the north end of Cariboo Lake.

A glacial moraine several hundred feet high lies immediately south of the canyon area. Natural features clearly indicate that part of a pre-glacial channel runs under the moraine. A slight depression suggests that its probable course is in a more or less straight line with the canyon.

The local geology also indicates promising bedrock values. Old workings in the vicinity, however, seem to have been directed at post-glacial concentrations on the south side of the creek (north of the pre-glacial channel). Coarse gold was found there.

On Pine Creek in the 1940s operations were recovering an average of 8/10ths of an ounce of gold per yard of washed gravel.

Between 1939 and 1945, with the Second World War in progress, very little mining was done. Aside from the lack of manpower, gold was deemed a non-essential mineral during the war.

Rich gravel deposits remain around Quesnel Forks because highgrading was accepted practice when natural conditions weren't just right. Miners required sufficient water pressure to process gravel through sluices and a grade of at least four per cent to wash away the tailings. Even then, gold values had to justify mining costs!

HARVEYS CREEK ▲

In 1932, prospectors found coarse gold on a gravel bench 25 feet above the creek - on the right bank a short distance upstream from the falls. High values were reported.

TWO MILE CREEK ▲

A small creek that flows into Cariboo Lake from the west two miles south of Keithley Creek. (It has its source in a small lake known as Two Mile Flat, which extends between Rollie (Duck) Creek and Keithley Creek at an elevation of 350 to 500 feet above Cariboo Lake.) An old channel cuts through Two Mile Creek 1.5 miles upstream from its mouth. Segments lie in both banks. The channel is about 15 feet wide and the downstream section lies in the creek's right bank. Surface testing here produced good results.

These last four creeks are located east of Horsefly and are not on our map, but they are on the same topographic map for Quesnel Lake.

ANTOINE (SUCKER) CREEK ▲▲

Flows into the Horsefly River three miles east of Horsefly. In 1929, gold was discovered half a mile upstream from Antoine Creek's mouth, immediately above a canyon. There's a trail in and recent sluicing proved the right bank is productive ground at least a mile upstream from here.

In 1896, the Horsefly River was producing well.

MacKay River ▲

Away south of Horsefly Lake, the MacKay River carries gold on all of its bars. At the turn of the century, when the MacKay was known as the South Fork of the Horsefly River, good prospecting is recorded from the mouth of Campbell Creek upstream.

Eureka Creek ▲▲

Eureka Creek was discovered near the headwaters of the South Fork of the Horsefly River (MacKay) in 1902. Panning produced an ounce a day per man and good paying gravel was being worked on the south fork as well at the time. This area has never been mined.

Frasergold Creek ▲▲

The discovery of float quartz in Frasergold Creek in the early 1900s suggests that gold washed into the creek from veins in the immediate area. Eureka, Empire (a tributary of Eureka) and Frasergold Creeks are all indicated in the records as relatively shallow - only two or three feet to bedrock.

From the 1902 records: "Upon reaching Eureka Creek, early miners fell short of provisions. They felled a tree and hollowed it out to serve as a sluice box - three inches deep and eight inches wide. Shoveling into this improvised sluice, they recovered about 1.5 ounces for each person per day of rough, coarse gold about the size of flaxseed. No doubt a lot of gold escaped due to the crude manner of the improvised box."

LIGHTNING CREEK

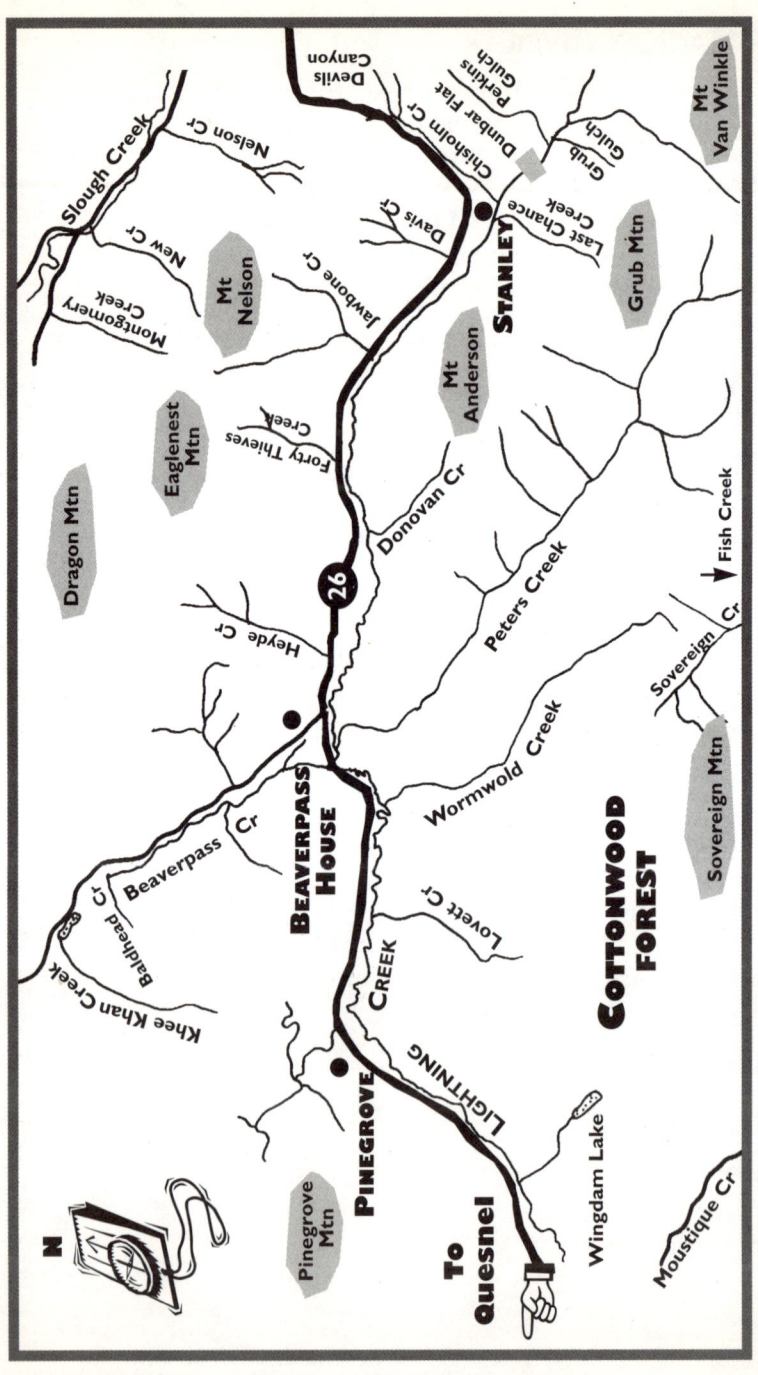

You been lookin' and prospectin' as the days stretch into weeks
You're scanning all the gravel bars and panning all the creeks
You're searching every riffle for the treasure each man seeks
In the streams that send their waters to the Fraser
- from The Mighty Fraser, by Roger Coe

CHAPTER 4

*L*IGHTNING CREEK

With its headwaters on Mount Agnes, Lightning Creek runs parallel to the Quesnel to Barkerville road (Highway 26) for some distance. The creek and its tributaries have produced gold from the 1860s on and the area continues to be rich in possibility.

DUNBAR FLAT ▲▲

In 1896, coarse gold was found on Lightning Creek just above Dunbar Flat near the former settlement of Stanley. It's likely this is an old portion of the Dunbar channel. Working just upstream from the flat and on the higher banks in the area, I've had excellent results with a gold pan and suction dredge. There are good possibilities too in a bench which lies 10 feet above the old worked channel. This bench is about 100 feet wide and 1,000 feet long.

CHISHOLM CREEK ▲

In 1878, gold was recovered on Lower Lightning Creek about four miles from its confluence with the Swift (or Cottonwood) River.

A small stream that flows into Lightning Creek at Stanley, Chisholm Creek's present course is recent in geological terms. The creek has an east and west branch (called Oregon Gulch), and it's believed an old channel runs into this gulch from the south and deposited the auriferous gravels in the area. The channel, now partially covered by a hill or slide area identifiable to the south of Chisholm Creek, should be examined. In 1915, Chisholm was producing nuggets weighing as much as an ounce. The tributary streams flowing in here have produced a lot of gold over the years.

MOUSTIQUE CREEK ▲

Point Claim Drift Mine, Lightning Creek 1900

Well worth exploring. Moustique Creek has a big, pre-glacial channel in the left bank three-quarters of a mile upstream

from its mouth on Lightning Creek. Look for exposed indications of the channel on both rims. This area is noted for coarse nuggets, but fine gold has also been panned here. Gold was first discovered just above the gorge, and the nuggets were found in cracks and crevices as well as in gravels.

Moustique Creek produced good results in 1910. Other creeks being worked at that time included Perkins Gulch, Last Chance Creek and Little Valley Creek.

Fish Creek ▲▲

A tributary of Sovereign Creek, Fish Creek appears to contain an old buried channel which could be an upstream extension of the Moustique Creek channel. Lovette Creek was also likely enriched by this same channel.

In the 1880s, a large part of Lightning Creek was being held by absentee owners and was not being worked.

Perkins Gulch ▲▲

Another tributary of Lightning Creek that displays strong indications of a productive old channel. I've been well rewarded prospecting through here.

Last Chance Creek ▲▲

A mile up from its source, Last Chance Creek has produced nuggets weighing more than an ounce. A pre-glacial channel lies in the right bank of this creek. In 1932 superficial diggings clearly showed that the modern creek cuts into the old channel at certain points. The rich returns suggested that gold from the ancient channel was being reconcentrated on the bedrock of the modern creek. Working this area in more recent times provided glittering proof of ongoing deposits.

Summit Creek was discovered in 1895 and the gravels proved rich in gold. Peters Creek exploration and mining also proved productive.

LOVERS' LEAP ▲

Near Lover's Leap (marked by provincial sign on the Quesnel to Barkerville road) a large bench area is still intact. There's evidence of a high channel crossing Lightning Creek in a northwest-southeast direction here. The channel's continuation northwest of the road is clearly defined by a valley of meadows. In a gulch on the right bank of Lightning Creek - immediately adjacent to the road and about 150 feet down - fairly coarse gold was found in glacial gravels overlying rock. Lack of water for hydraulic mining in the early 1930s hindered work here.

DEVILS CANYON ▲

Small operations were worked on Devils Canyon (Lake) Creek about five miles west of the town of Wells in the 1940s.

Farther east on the Quesnel to Barkerville road, Devils Canyon was the site of a new discovery in 1938. Meadowlands are located on both sides of the Chisholm Creek / Devils Lake Divide and the discovery was made at a meadow on the northend of the canyon, upstream from old workings. Earlier miners had concentrated their efforts on the post-glacial deposits above the east wall of Devil's Canyon.

BEAVER PASS ▲

In 1949, preparations were being made to hydraulic mine the Devils Canyon channel, which cuts around the lower stretch of Burns Creek.

A wide valley, Beaver Pass deserves close inspection of both its tributaries in the pass and of the region beyond the north end of Ahbau Lake. For years gold was recovered from the creeks around here. Good returns obtained by prospectors working by hand in 1932 sparked renewed

**1870s prospector
tending dumpbox,
Lightning Creek.**

interest in the area. There are known
concentrations of gold on the bedrock
and false bedrock of the creeks - many of
which occupy post-glacial channels.
The streambeds cross pre-glacial channels,
which in several places are plainly indicated,
lying buried on one side or the other of
the existing creeks.

BALDHEAD AND KHEE KHAN CREEKS ▲▲

Owing to the general hard times and scarcity of other work during the First World War, many placer miners were forced back to creeks the 'Old Timers' said were worked out. Luckily, though, many of these so-called, worked-out creeks proved once again to be lucrative.

A similar situation exists along Baldhead Creek and Khee Khan Creek, both of which run into Beaver Pass just north of Four Mile Lake. These creeks have cut through old channels of gold-bearing gravel and have reconcentrated the gold on their creek beds. It seems likely that some of these channels are native to the Beaver Pass (Bedrock) valley and that the placer material is of local origin. At one point, Baldhead Creek flows along bedrock on the right rim of an ancient channel. This channel is buried in the left bank of the creek and the area has produced fairly coarse, well worn gold.

NO NAME CREEK ▲ (not on map)

Ground sluicing was done in the 1930s on Kwong Foo Creek, which runs parallel to Slade Creek about 1,000 feet away.

In 1938, a new discovery was made 3.5 miles up from Beaver Pass House. Coarse gold on true bedrock was found a few feet below clay on No Name Creek, which flows into Beaver Pass from the northeast, north of Baldhead Creek.

GAGEN CREEK ▲ (not on map)

Flows into Lightning Creek from the south, near Coldspring House. At 1.5 miles upstream from the confluence, a new discovery in 1932 turned up spectacular gold in the right bank of Gagen Creek. Look for an area where rock appears near the creek's surface and frequent outcrops are seen. The bench ground along here is only a few feet above the creek and was so rich that 37 ounces of gold were obtained

from 275 yards of gravel. Below this point
the creek has deepened its bed and the
benches for half a mile run to a width of
200 feet, offering good possibilities. The
higher banks in this area are also productive. In recent times, I've found coarse
gold, fairly worn, in cracks and crevices as
well as in gravels. Gagen Creek eventually
enters Lost Valley and flows northwest to
its junction with Lightning Creek.

Two Channels ▲▲

Two channels cross Lightning Creek in
the bench region just described. These
channels are roughly three-quarters of a
mile apart. One is the valley through
which the upper portion of Gagen Creek
flows, the other is Lost Valley. Lost Valley
is dry, but is clearly defined and more than
600 yards wide nearing Lightning Creek.

RIVALRIES

"There was this one chap, he always had the great habit of telling me that he was going to stake me in. Meaning, that if I was staking mining claims, he was gonna come along and by the time I got my first claim done he would have all the rest in the area staked.

And he was always buggin' me about this. So it became ... not exactly adversarial, but it became sort of a: 'Well if you're in an area I know it's good, so I'm gonna be in there too and I'm gonna get anything I can before you get it all' - sort of thing, eh?

And this went on for a few years.

So I took one of my prospecting partners - this was in December and it was getting close to Christmas ... matter of fact I think it was about the 21st of December ... anyway, we knew these claims were coming open at midnight tonight. This was the anniversary date.

So we went out at about 8 o'clock figuring we'd be on the ground at midnight. This used to be a favourite trick years ago - go in at midnight and start your staking and by 9 o'clock in the morning you're in the recorder's office recording the claim!

We took a third guy with us. The other chap that was always with me, he had bought himself four bottles of rye for Christmas and I think a bottle of gin or something. It was likely his liquor supply for the whole winter 'cause he didn't drink that much. So we drove out and we came to a fork in the road and all of a sudden here's vehicles tracks ahead of us in the fresh snow.

Well I knew there was a cabin just down the road from the claims that we wanted. 'We're in trouble now,' I said. 'There's somebody else in here probably got the same idea as us, gonna stake these at midnight.'

So we drove in a little ways further and when I got about a quarter of a mile away from the cabin I shut the truck off. I'd shut the lights off before that because, with the snow, there was enough visibility to see. We get up close to the cabin ... and I see a red Ford there which I recognize immediately as belonging to this guy that was always buggin' me and saying he was gonna stake me in.

So I walked back to the pick-up and told the guys what was going on. And I'm trying to think of what I can do, because if we drive by with our 4x4 they're gonna spot us immediately.

So I said to the older chap: 'Can I borrow two bottles of rye from you ... until tomorrow, when we go back to town?' He looked at me like ... you know ... what did I want two bottles of rye for? 'Well I'd like to borrow them until tomorrow,' I said. 'I'll replace them in the morning.'

"Well okay,' he said.

Then I turned to the third chap - and this third one he was a real drunk, he really got into it - and I said to him: 'I want you to take your rifle and packsack and knock on the door of that cabin and tell 'em you've been hunting, that you stayed out too long in the bush and that you'd like to stay the night .... Produce one bottle of rye,' I said, 'but don't produce both of them immediately.'

And I told him to be on the road the next morning by 7 o'clock otherwise he'd be walking out. And I said: 'Now don't you go and get loaded. Make sure you give them more than you take yourself when you pour the drinks. And you pour it! You load their cups up good and keep yours light.'

Well, this worked. When the other chap and I drove by at about 11 o'clock that night we could hear the laughing and the hootin' and the hollerin,' and we went in and we staked the claims.

Next morning I see this chap and he's sittin' on the side of the road with his head in his hands. "Jeez, I sure could use a drink,' he says.

Well, we had about a 16 mile drive to town and we got in too early for him because the beer parlours weren't open yet or anything and we had to sit around for a while. But as soon as the Mining Recorder opened I was in there recordin' those claims immediately.

About a week later, I ran into the two guys from the cabin on the street. 'How's the staking in business?' I said.

And the one guy that was always buggin' me, he said: 'You dirty sonofabitch. You sent him in there to get us pissed didn't you?'

And I said: 'I didn't do anything.'

And he said: 'No, no, no! ..... We figured it out ... We followed his trail out and we saw where he got back in your pick-up.'

And I said: 'Well, that's one time you got staked in!'

Well, they didn't like it. I mean, I guess they were pretty angry the next morning, but they just took it all in good stride.

It just goes to show ..."

- Jim Lewis

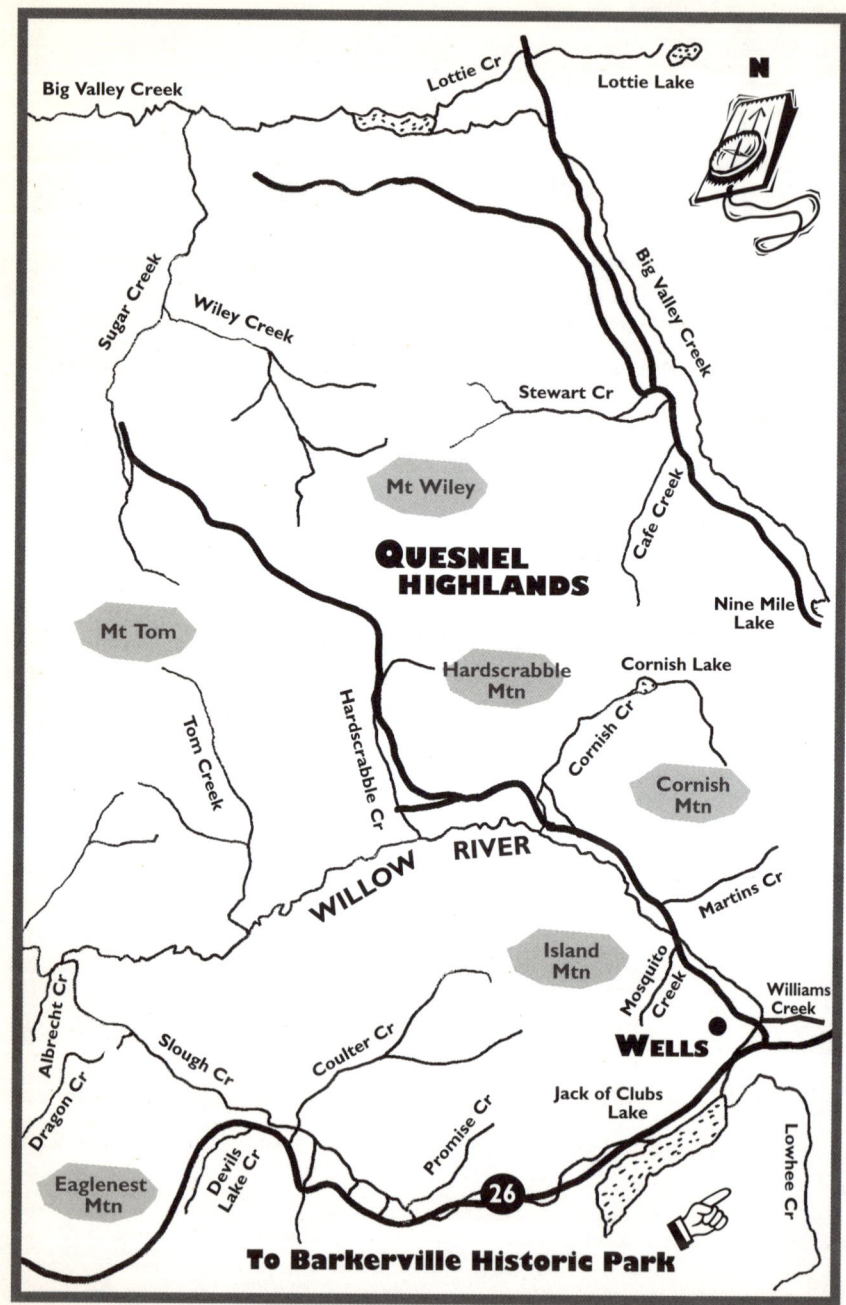

WILLOW RIVER

*J*f you felt the evenin' breeze rustle through the autumn leaves,
You've had gold dust sprinkled through your fingers and your mind.
It's all here for me and you, the great 'Chilcotin-Cariboo'
So come and play a part in Mother Nature's splendid game.

- *from Go for the Gold*, by Jean Ann Robitaille

CHAPTER 5

*W*ILLOW RIVER

The Willow River, renowned for deep ground and buried channels, drains the largest area of rich placer ground in the Cariboo. Williams Creek - which runs through the heart of historic Barkerville - and Jack of Clubs Creek together form the river's east fork. Both merge with the Willow River at Wells. The river's west fork, once known as Lost Creek, is now named Slough Creek. Its confluence with the river lies seven miles west of Wells on the opposite side of Island Mountain.

The Willow River has immense gravel banks and almost every tributary ever prospected has produced paying gold, especially the creeks entering from the east side. Still, opportunities abound beyond the sites mentioned: on Hardscrabble Creek, Sugar Creek, Tregillus Creek and Rucheon Creek.

BIG VALLEY CREEK ▲

Lying northwest of Barkerville, Big Valley Creek has its source in Nine Mile Lake. The creek runs for miles through a broad valley that ranges from 500 to 1,000 feet wide. Plenty of gold has been found close to the valley's surface and it still shows excellent potential. There's a canyon five miles downstream from the source, below which the valley narrows to 400 feet. Below the canyon, miners were

In 1876, two companies were prospecting Carry On Creek, a tributary of the Willow River.

Café Creek was being prospected and developed for mining in 1903. Stewart Creek was averaging 2.5 to 3 ounces of gold for every 10 feet worked.

enjoying good prospects at the turn of the century, but work halted abruptly in the face of high water. The tributaries flowing in from the south - Café, Stewart and Sugar Creeks - have all been productive and there's no question that an immense amount of gold-bearing gravel is located in this area.

DEVILS LAKE CREEK ▲

Another area that was never worked thoroughly in earlier times due to lack of water for hydraulic mining. Past records suggest there were two distinct runs of gold - one running up Devils Lake Creek, the other running roughly parallel to Slough Creek.

Miners celebrate Christmas at Dead Lead Camp, Slough Creek 1903

SLOUGH CREEK ▲▲

Slough Creek is a relatively small stream running through a flat-bottomed

valley. From along its southern tributaries - Nelson, Burns and New creeks - a lot of gold was taken in the early 1900s. A bench of paying gravel extends along the south side of the valley and an old channel runs south towards Lightning Creek.

MOSQUITO CREEK ▲

Not to be confused with Moustique Creek, Mosquito Creek is rather unusual: gold values have been found higher up - on the rim rock - rather than on bedrock. Although it's a short creek that flows into the Willow River only 1.5 miles northwest of Wells, plenty of paying ground had still not been worked by 1914, despite the rich gravels.

ALBRECHT CREEK ▲▲

Located two miles downstream from Dragon Creek, Albrecht Creek has an intriguing history. In the early 1900s, a prospecting tunnel was driven in to the banks to tap into a channel that was previously discovered in the hillside. The principal owner of the only area being worked by 1906, however, was struck by a fatal illness and the property was shutdown - in spite of the fact that results were improving as the work progressed.

DRAGON CREEK ▲▲

An old channel crosses Dragon Creek half a mile upstream from its mouth. The channel parallels the Willow River for 16 miles, running in a slightly more westerly

Burns Creek, Slough Creek, Dragon Creek and Coulter Creek were being worked on a steady basis, but conditions were dry and hydraulic and sluice operations required massive amounts of water to wash the high banks.

On McCarters Gulch, Conklin Gulch and Hardscrabble Creek, prospecting was good in 1899. Another tributary, Cornish Creek, produced gold in shallow ground near its source.

In 1901, mining tunnels were driven in with some success, but flooding kept driving back the miners.

Northeast of Barkerville, the Bear and Goat rivers were prospected and hold great promise.

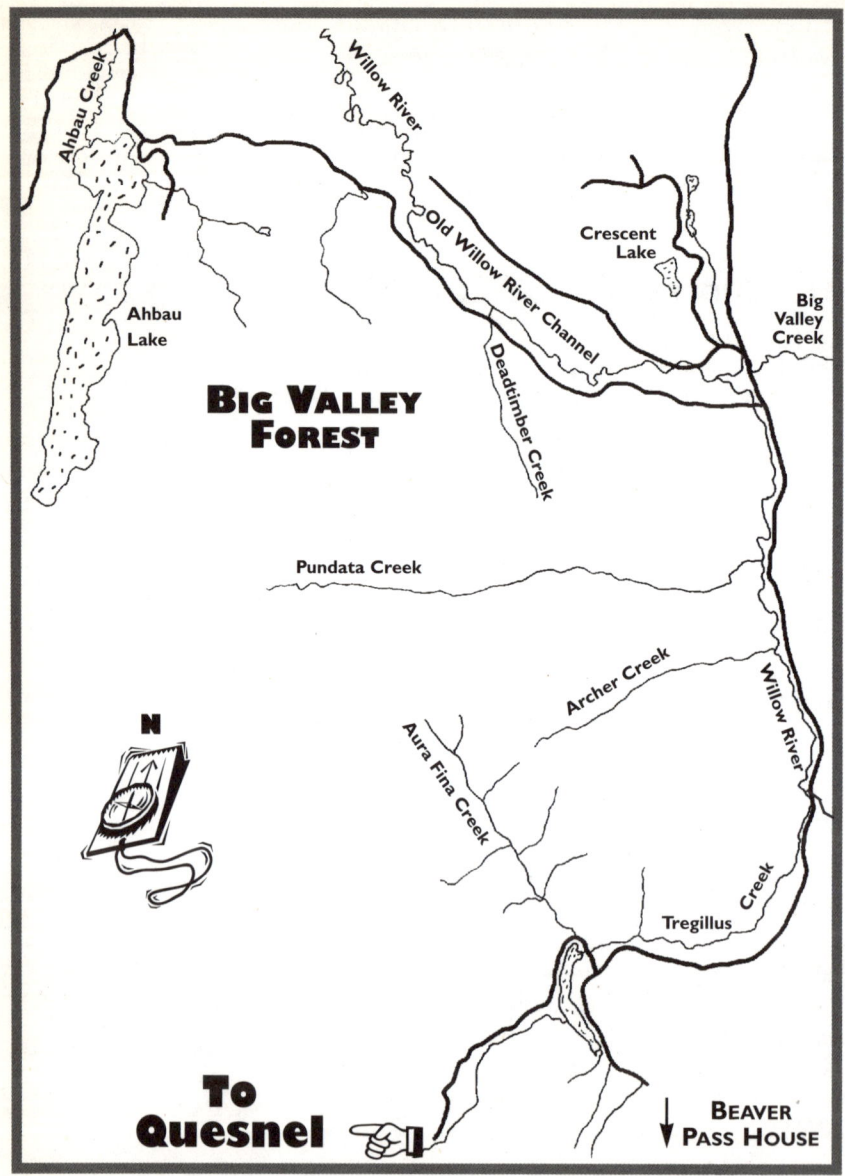

WILLOW RIVER
(northeast of page 76 map)

direction, before turning east and meeting the river. This is believed to be the old channel of Lost Creek (later renamed Slough Creek). From knowledge gained at the Dragon Creek exposure, and from other prospecting evidence, it's believed this channel crosses Tregillus Creek. The channel is quite likely the source of gold found in Rucheon and Baldhead creeks, as well as the rich spots found in the right bank of Tregillus Creek. From here to the point 10 miles farther north where it meets the Willow River, the old channel is well defined.

Larsen Gulch, off Rucheon Creek, was worked in 1943 with pleasing results for a small scale operation.

ARCHER AND DEADTIMBER CREEKS ▲▲

Archer and Deadtimber creeks empty into the Willow River north of Tregillus Creek (named after the family pictured on page 46) and cut through the same channel. Both creeks have been worked successfully upstream and it's more than likely that the old channel is the source of their gold.

AHBAU CREEK ▲

Located between Lodi Lake and Ahbau Lake, Ahbau Creek is still productive, even though it has surrendered plenty of gold over the years. All the creeks draining from the west into Ahbau Creek have paid good dividends from surface work. Judging from recent prospecting in the area, it is still promising.

Shepherd Creek, Hardscrabble Creek and Grouse Creek were all giving good returns for the amount of work done. "Each year, high water rushes down more auriferous gravels and more gold is deposited in the stream beds. People have been known to profitably work the same portion of the creek bed for 10 years or more."

Aura Fina Creek ▲▲

In 1951 work started on a new area above the canyon on Aura Fina Creek after evidence of an ancient channel was discovered on the west bank. The canyon is 1.5 miles upstream from Tregillus Creek.

Old Willow River Channel ▲▲

In 1941 labour for mining was frozen, gold being a non-essential war mineral. Consequently placer mining slowed to a minimum during the war years.

The Willow River in pre-glacial times used to occupy a channel a mile or more east of its present position. Glacial drift (the George Creek glacier) blocked this former channel and forced the river into several different courses to the west until, eventually, it cut the bed it now occupies. The pre-glacial channel lies mainly to the east and the rock benches there expose the successive channels occupied by the river.

Lowhee Creek & Mine ▲

Until it shut down in 1947, the Lowhee Mine was the oldest continuously operated

The Mucho Oro Goldmining Company, Lowhee Creek 1868

placer mine in the Cariboo. Lowhee Creek has been extremely productive in the past, yet ground was left untouched between the mine and Stouts Gulch.

COULTER CREEK ▲

Try working a mile upstream from Coulter Creek's mouth. Look for signs of a buried channel on the north side about 30 feet above creek level.

Despite encouraging returns from prospecting on Coulter Creek, no serious mining was undertaken. The Eight Mile Lake area also showed promise in the creek beds.

WILLIAMS CREEK ▲

As early as 1876, an area just above Mink Gulch was being successfully mined, although more attention was being paid to the east bank of Williams Creek. In the early 1900s, a blind channel about 40 feet wide was discovered and its walls were recorded as 30 feet high. Despite intense interest from the onset of the Cariboo Gold Rush, in 1932 a productive new portion of the creek was discovered opposite Mink Gulch. Williams Creek is today still attracting lots of interest from small companies and individual prospectors.

1904 records say that Williams Creek had been continuously worked for 44 years - but only over a two and a half mile section.

STOUTS GULCH ▲

Very rich ground, again overlooked by previous miners, was found on Stouts Gulch in 1913. A tributary of Stouts Gulch - Emory Gulch - also produced excellent results at that time. These last two examples, both in popular areas that are now part of a historic park, illustrate how easy it is to miss paying ground - even when it's right under your nose!

In 1952 Conklin Gulch and Emory Gulch, a tributary of Stouts Gulch, were hydraulic mined in a small way.

Horsin' around

" 'You killed my horse!'

The sudden sound of a stranger's voice behind me as I was panning along a creek, lost in my own thoughts, shocked the hell out of me.

I turned and saw an old fella standing over on the far shore looking at me. There weren't any high banks or anything and his words rang in my ears: 'You killed my horse.'

The old boy only lived maybe 75 feet away from the creek. He had a little trail leading in there, but I never saw it because I wasn't paying any attention to that. I noticed he had a limp as I followed him along this trail.

'You shot my horse,' he said.

'Your horse?!?'

We arrived in no time at all outside a small cabin. The thing showed he was ingenious enough. He'd dug himself a place in the bank and then built a framework out on the front of it. I didn't go inside, but from what I saw, the logs of the cabin only went out about four or five feet and the rest of it was dug into the ground. He had a stove pipe there and he was a bit of a recluse.

He led me over to the corral and I got another shock. Lying on the ground there was the bleached bones of a horse!

We talked about it and after a little while he said: 'Ah, you're a pretty good fella. I guess you didn't kill my horse.'

And I said: 'No, I didn't. It looks like it's been dead quite a while.'

'No, no, no,' he said. 'It's only been about a week or so. Somebody killed it on me.'

What had happened, I found out from some people later, was he fell and he broke his leg. Apparently he was laid up for quite a while inside the cabin. He couldn't get out to feed the horse and it starved to death in the corral. There was no hay in the winter time.

I guess the old guy might have bumped his head too when he broke his leg. I don't know what happened to him exactly, but he was a bit touched, all right. Apparently some local people used to take care of him after that. About once a month or so they'd go up there and make sure he was okay.

But that damn horse was just laying there, you know. It was just a skeleton!"

- Jim Lewis

HIXON CREEK

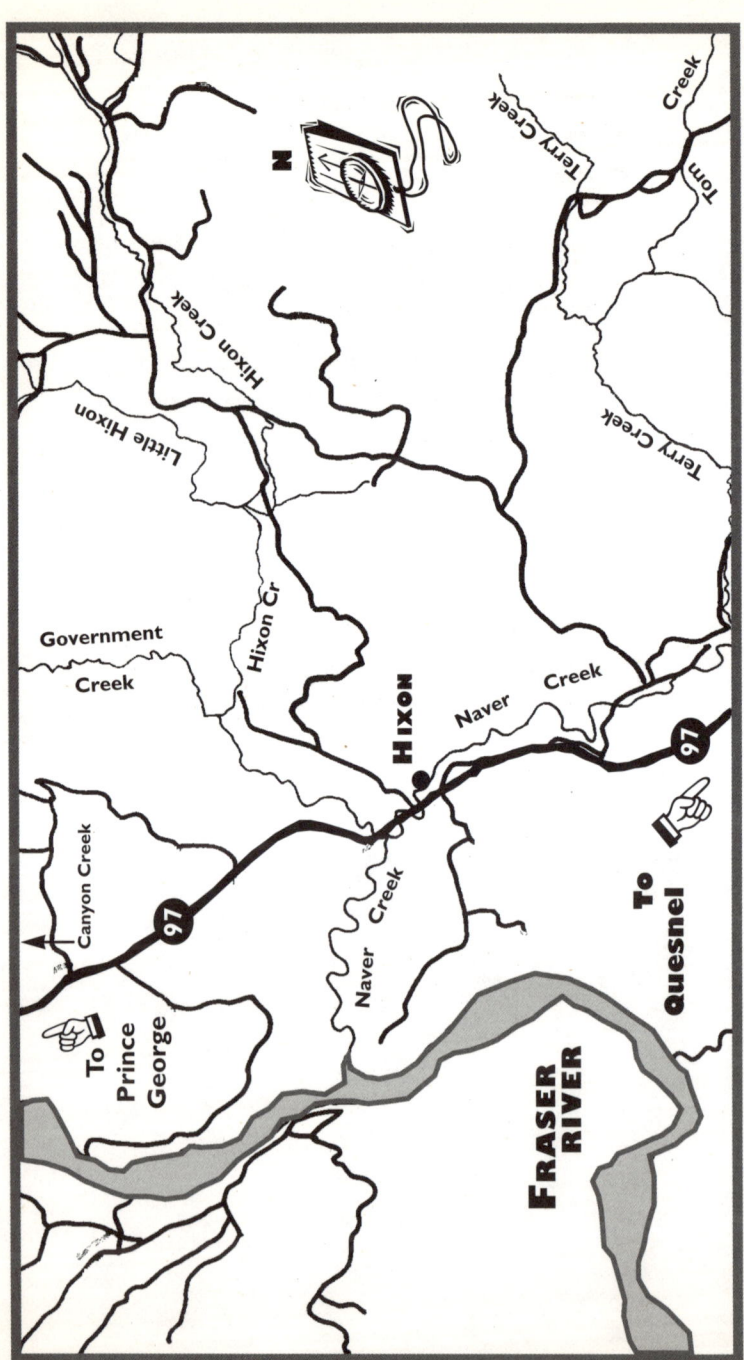

When you're sittin' by your fireside some winter's night so cold
And the talk turns to adventure, to men both brave and bold
You will tell them of the Cariboo's wild stampede for gold
And you'll tell them of the roarin', rollin' Fraser.

- *from The Mighty Fraser, by Roger Koe*

CHAPTER 6

*H*IXON CREEK

Hixon Creek joins the Fraser River about 35 miles south of Prince George. First prospected in the late 1800s, exploration of placer ground began just before the turn of the century. Records indicate that by 1935, none of the pre-glacial channels on Hixon Creek had been touched. The next mention of Hixon Creek is 1948, when only one lease was under production. Since then, surprising as it may seem, there's been little recorded interest in the area.

BEDROCK VALUES ▲▲

Hixon Creek gravels overlay a band of white clay which can easily be mistaken for bedrock. Don't be fooled: coarse gold has been found in this clay to a depth of several feet. Along Hixon Creek and its tributaries, good paying gold has been found on bedrock, on false bedrock and in the high benches. Just downstream from its confluence with Government Creek, placer gold was found right on the surface of Hixon Creek in 1922.

HIXON CREEK FALLS ▲

Three quarters of a mile upstream from its confluence with Government Creek,

In 1902 there were seven men working on Hixon Creek. A small amount of work was done on Government Creek below the Buckley Creek junction.

Very nice gold was being recovered on this creek in 1930, but very little work was carried out.

Chinese-built pipeline and trestle, Canyon Creek 1897

90-foot falls descend in three steps. Beyond the falls for several miles upstream wire gold veins in quartz pieces have been found. Within an area of 40 square feet miners recovered $500 in gold when prices were $17 per ounce. They only worked the surface to a depth of three feet. I've prospected claims here fairly recently (for their owners) and the results were excellent. Mining records for 1927 indicate that a large bench - 450 feet long and 300 feet wide - is located farther upstream and at that time had still to be prospected.

OLD CHANNEL ▲▲

Upstream from the falls, a pre-glacial channel appears to run along the north side of Hixon Creek for about 3.5 miles. The channel eventually cuts across to the south side of the creek somewhere below the falls.

Low Bench ▲

In 1931, a spectacular new discovery was made on Hixon Creek. You can see that most of the creek's left bank above the falls sits on a low-lying bench. Miners recovered more than 74 ounces of gold from only a small yardage of gravel here. And they did it while shovelling gravel over their heads to reach the height necessary to work the sluices!

In 1949 and 1950, only one lease was worked, the balance of the area being largely ignored.

Government Creek ▲▲

Along with its tributaries, this area is highly recommended. The whole watershed, which follows a southerly course to its confluence with Hixon Creek, was overlooked for many years. In 1915, enterprising prospectors at work on Government Creek found coarse, heavy gold and nuggets weighing more than an ounce. Similar gold values have since been found two miles upstream from the confluence, at a point where gravel and clay banks rise 100 to 150 feet above creek level. Although some of the gold found here was fine, there was almost no evidence of flour gold. This suggests that the gold originated locally. I've prospected this area and found that black sand is also present, along with limited values in platinum.

Starting in the 1950s, frequent conflict arose between mineral title holders and placer claim holders. This limited exploration of the area.

Canyon Creek ▲▲

A large stream that follows a wide valley west until, 20 miles upstream, it runs through a canyon for five miles. In 1930, gold was discovered at the lower end of the canyon where the valley opens out

In 1954, one miner recorded a small amount of sluicing on Government Creek; on Hixon Creek, only one lease area was being worked.

again. Coarse nuggets turned up in the river bed and gravel banks near a bench roughly 500-feet wide. The low benches on both sides of the creek also held coarse, placer gold. From here the creek runs north in a wide, flat-bottomed valley until it reaches the Fraser River.

LITTLE HIXON CREEK ▲

Slides and floods stopped most work in 1962; by 1965 work had resumed but was soon halted due to a shortage of water.

Also known as the north fork of Hixon Creek, Little Hixon Creek is well worth exploring. In 1932, very coarse gold and nuggets weighing as much as four ounces were found 400 yards upstream from its confluence with Hixon Creek. A shortage of water for sluicing and subsequent legal problems suspended work on the creek. I've prospected here and it's likely an ancient channel segment is located in this area.

TERRY CREEK ▲▲

A 1967 slide of overburden into the mining area stopped the recovery of gold until the debris could be cleared.

Coarse gold and large quartz rocks have been found on Terry Creek. Upstream, immediately above Tom Creek, low-lying benches of auriferous gravels line both banks. The valley here is between 500 and 900 feet wide. Farther on, the creek runs through a rocky gorge for some distance. Three miles upstream from Tom Creek, it enters a canyon for 300 yards, above which the valley widens again. At its upper end, the canyon's walls are 90 feet high and a clearly defined channel runs northwest toward Hixon Creek.

A commemorative
cairn marks the site
of the Cariboo Gold
Field, Barkerville
1959

GOLD COMMISSIONER OFFICES

Cariboo Mining Region
102 - 350 Barlow Street
Quesnel, BC V2J 2C1
Ph: (250) 992-4310
Fax: (250) 992-4314

Coast/Liard Mining Region
302 - 865 Hornby Street
Vancouver, BC V6Z 2C5
Ph: (250) 660-2669
Fax: (250) 660-2653

Kamloops/Okanagan Mining Region
250 - 455 Columbia Street
Kamloops, BC V2C 6K4
Ph: (250) 828-4544
Fax: (250) 828-4542

Coast/Liard Mining Region
302 - 865 Hornby Street
Vancouver, BC V6Z 2C5
Ph: (250) 660-2669
Fax: (250) 660-2653

Kootenay Mining Region
310 Ward Street
Nelson, BC V1L 5S4
Ph: (250) 354-6109
Fax: (250) 354-6407

East Kootenay Mining Region
100 Cranbrook Street North
Cranbrook, BC V1C 3P9
Ph: (250) 426-1249
Fax: (250) 426-1252

Omineca Mining Region
1020 Murray Street, Bag 5000
Smithers, BC V0J 2N0
Ph: (250) 847-7207
Fax: (250) 847-7232

Vancouver Island Mining Region
3rd Floor-1810 Blanshard Street
Victoria, BC V8V 1X4
Ph: (250) 952-0567
Fax: (250) 952-0541

ASSISTANT GOLD COMMISSIONER OFFICES

Alberni Mining Division
4586 Victoria Quay
Port Alberni, BC V9Y 6G3
Ph: (250) 724-9200
Fax: (250) 724-9298

Atlin Mining Division
Box 100, 3rd Street
Atlin, BC V0W 1A0
Ph: (250) 651-7595
Fax: (250) 651-7707

Clinton Mining Division
1423 Cariboo Highway, Box 70
Clinton, BC V0K 1K0
Ph: (250) 459-2268
Fax: (250) 459-7922

Golden Mining Division
903-9th Street South, Box 39
Golden BC, BC V08 1H0
Ph: (250) 344-7552
Fax: (250) 344-7553

Greenwood Mining Division
524 Central Avenue, Box 850
Grand Forks, BC V0H 1H0
Ph: (250) 442-5444
Fax: (250) 256-4546

Lillooet Mining Division
651 Main Street, Bag 700
Lillooet, BC V0H 1H0
Ph: (250) 256-7548
Fax: (250) 256-4546

Nanaimo Mining Division
13 Victoria Crescent
Nanaimo, BC V9R 5B9
Ph: (250) 741-3636
Fax: (250) 741-3622

**New Westminster
Mining Division**
100-635 Columbia Street
New Westminster, BC V3M 1A7
Ph: (250) 660-8672
Fax: (250) 660-8498

Nicola Mining Division
2176 Quilchena Ave., Bag 4400
Merritt, BC V0K 2B0
Ph: (250) 378-9343
Fax: (250) 378-9346

Osoyoos Mining Division
112-100 Main Street
Penticton, BC V2A 5A5
Ph: (250) 492-1211
Fax: (250) 492-1213

Revelstoke Mining Division
104-1123 2nd Street West
Revelstoke, BC V0E 2S0
Ph: (250) 837-7635
Fax: (250) 837-7640

Similkameen Mining Division
151 Vermillion Avenue, Box 9
Princeton, BC V0W 1W0
Ph: (250) 295-6957
Fax: (250) 295-3070

Skeena Mining Division
100 Market Place
Prince Rupert, BC V8J 1B8
Ph: (250) 624-7415
Fax: (250) 624-7421

Slocan Mining Division
312-4th Street, Box 580
Kaslo, BC V0G 1M0
Ph: (250) 353-2219
Fax: (250) 353-2316

Trail Creek Mining Division
1050 Eldorado Street
Trail, BC V1R 3V7
Ph: (250) 364-0591
Fax: (250) 364-0561

Vernon Mining Division
3201-30th Street
Vernon, BC V1T 9G3
Ph: (250) 549-5511
Fax: (250) 549-5508

Maps - BC
Ministry of Environment, Lands and Parks, Geographic Data BC
4th Floor, 1802 Douglas Street, Victoria, BC V8V 1X4
Ph: (250) 387-1441 Fax: (250) 387-3022
Website: www.env.gov.bc.ca

CREEK INDEX

ABOUT THE AUTHORS

Born in New Brunswick, Jim Lewis moved west in the early 1950s to prospect for gold. He spent the next 40 years staking claims all over BC and the Yukon - walking, canoeing, four-wheeling and flying into remote places. In that time, he has also owned and operated several B.C. placer mines.

With this wealth of experience, there isn't a gold trick that Jim isn't wise to . . . or willing to discuss.

'Of all the gold in the world, only .02 per cent has ever been mined,' he says, a knowing twinkle in his eye. Jim is confident that anyone who knows where to look will discover for themselves the excitement of panning for gold.

Now semi-retired and living near Nanaimo, B.C., Jim Lewis is a consultant for both placer and hardrock mining.

Born in Montreal and raised in England, Charles (Chuck) Hart now calls Vancouver Island home.

As a musician and erstwhile treeplanter he has worked and played in communities (and the bush) throughout B.C. - fiddling on the streets of Barkerville and performing songs about the Cariboo gold rush.

A journalist, writer and editor with wide-ranging interests, Chuck's fingers seldom lie still. Away from the keyboard, they can usually be found tinkering on a musical instrument.

MORE BOOKS ON GOLD

Methods of Placer Mining
by Garnet Basque. Explains the techniques used to recover gold and illustrates equipment and how to build a sluice box. Reviews the world's major gold rushes. $6.95

Lost Bonanzas of Western Canada
Outlaw loot, lost mines and sunken bullion are some of the lost bonanzas in Alberta, British Columbia and the Northwest Territories. Do these treasures exist or are they imaginary? $14.95

Gold Panner's Manual
by Garnet Basque. With over 190,000 copies in print across North America, this BC based book is an ideal complement to the Heritage Creeks of Gold series; highly illustrated. The ABC's of Goldpanning and Recovery. $12.95

Cariboo Gold Rush
The saga of 30,000 chasing one of the world's richest discoveries in 1858. They found nuggets by the ton and carved a web of trails and memories into the heart of BC. Maps, photos, bar room tales, success, tragedy and characters galore. $7.95

BC Recreational Atlas
This very popular travel companion, developed in conjunction with BC government sources is the most complete reference atlas available for backroaders, campers and general travel and is updated regularly. $23.95